The Lower Thames Valley is a classic area for British Pleistocene studies. It extends downstream from central London to Tilbury. The valley contains a sequence of River Thames deposits representing approximately the last 300 000 years. It includes older, highly fragmented and eroded sediments derived from Thames tributaries and glaciation. The Lower Thames includes some of the most important Palaeolithic archaeological sites in the country, such as Swanscombe, Northfleet, Grays, Crayford and Stoke Newington. These sites have been investigated in great detail in the past, but have never previously been fitted into a regional context. This study does this and considers the significance of the finds in relation to their geological setting. The area also includes some of the most important fossiliferous localities in the country, several of which have been at the centre of significant controversies regarding the sequence of events in the British Pleistocene. This regional investigation clarifies the problems by presenting the geological sequence in detail and establishing clearly the relationship of these important localities for the first time.

This fascinating account provides a companion study to *The Pleistocene History of the Middle Thames Valley* (by the same author) and will be of interest to those studying river and landscape history.

Pleistocene History of the Lower Thames Valley

Pleistocene History of the Lower Thames Valley

PHILIP L. GIBBARD

*Assistant Director of Research, Subdepartment of Quaternary Research,
Botany School, University of Cambridge*

CAMBRIDGE
UNIVERSITY PRESS

CAMBRIDGE
UNIVERSITY PRESS

University Printing House, Cambridge CB2 8BS, United Kingdom

One Liberty Plaza, 20th Floor, New York, NY 10006, USA

477 Williamstown Road, Port Melbourne, VIC 3207, Australia

4843/24, 2nd Floor, Ansari Road, Daryaganj, Delhi - 110002, India

79 Anson Road, #06-04/06, Singapore 079906

Cambridge University Press is part of the University of Cambridge.

It furthers the University's mission by disseminating knowledge in the pursuit of education, learning and research at the highest international levels of excellence.

www.cambridge.org
Information on this title: www.cambridge.org/9780521402095

© Cambridge University Press 1994

This publication is in copyright. Subject to statutory exception
and to the provisions of relevant collective licensing agreements,
no reproduction of any part may take place without the written
permission of Cambridge University Press.

First published 1994

A catalogue record for this publication is available from the British Library

Library of Congress Cataloging in Publication data
Gibbard, Philip L. (Philip L.), 1949–
Pleistocene History of the Lower Thames Valley / Philip L. Gibbard.
   p. cm.
Includes bibliographical references and index.
ISBN 0-521-40209-3
1. Geology, Stratigraphic–Pleistocene.   2. Glacial landforms–England–Thames River Valley.
3. Paleolithic period–Engand–Thames River Valley.   I. Title.
QE697.G386   1994
551.7'92'09422-dc20   93-31544   CIP

ISBN 9780521402095 Hardback

Cambridge University Press has no responsibility for the persistence or
accuracy of URLs for external or third-party internet websites referred to in
this publication, and does not guarantee that any content on such websites is,
or will remain, accurate or appropriate.

Contents

Preface   ix

1 Introduction   1

1.1 Topography and geological background   4
1.2 Previous investigations   4
1.3 Note on terrace stratigraphy   8
1.4 Nomenclature of terrace deposits   9

2 Lithostratigraphy   11

2.1 High-level ('pre-Thames') deposits   12
2.2 Dartford Heath Gravel   18
2.3 Swanscombe Member Deposits   24
2.4 Orsett Heath Gravel   29
2.5 Corbets Tey Gravel   34
2.6 Mucking Gravel   49
2.7 Aveley/West Thurrock/Crayford Silts and Sands   59
2.8 West Thurrock Gravel   89
2.9 Kempton Park/East Tilbury Marshes Gravel   89
2.10 Langley Silt Complex ('brickearth')   94
2.11 Shepperton Gravel ('Lower Floodplain' or 'buried channel' infill)   100
2.12 Tributary valley floodplain gravels   107
2.13 Staines Alluvial Deposits/Tilbury Deposits   114
2.14 Lea Valley floodplain alluvium   117

3 Comparison of the pebble lithological composition of the gravel members   119

3.1 Principal components analysis   119

4 Sedimentary structures and depositional environments   123

4.1 Gravel units   123
4.2 Mass flow deposits   126
4.3 Sand and fine sediment association   127

5 Vertebrate faunal assemblages   131

5.1 Taphonomy and preservation   134
5.2 Faunal assemblages   135

6 Palaeobotany and biostratigraphy — 140

- 6.1 Grays Thurrock — 140
- 6.2 West Thurrock — 143
- 6.3 Aveley, Sandy Lane Quarry — 145
- 6.4 Purfleet — 146
- 6.5 Belhus Park, Aveley — 146
- 6.6 North Ockendon — 149
- 6.7 Ilford — 149
- 6.8 Nightingale Estate, Hackney — 153
- 6.9 Highbury — 155
- 6.10 Peckham — 158

7 Palaeolithic artefact assemblages — 162

- 7.1 Incorporation of Palaeolithic artefacts in sediments — 163
- 7.2 Relationship of the artefacts to the stratigraphical units — 166

8 Palaeogeographical evolution of the Lower Thames Valley — 174

- 8.1 Epping Forest Formation (pre-Anglian–?pre-glacial Anglian) — 178
- 8.2 Warley Gravel and Darenth Wood Gravel (pre-Anglian) — 178
- 8.3 Anglian Stage — 178
- 8.4 Hoxnian–early Wolstonian Stages — 181
- 8.5 Wolstonian Stage — 183
- 8.6 Ipswichian Stage — 185
- 8.7 Devensian Stage — 191
- 8.8 Flandrian Stage — 193

9 Correlation with neighbouring areas — 195

- 9.1 Central Essex — 195
- 9.2 Middle Thames Valley — 197
- 9.3 South-eastern Essex — 198

References — 201

Appendix 1 Pebble counts from high-level gravels in the Epping Forest area — 214

Appendix 2 Pebble counts from the Lower Thames region — 216

Index — 222

Preface

The Thames Valley is considered by many to be one of the classic regions in British Pleistocene geology. This is because it contains the longest and potentially the most complete sequence of Pleistocene fluvial and associated sediments outside East Anglia. The Lower Thames Valley, east of central London, is extremely well known for its abundance of fossiliferous localities, the age and significance of which have been the subject of controversy, particularly during recent years. However, although a number of these sites or small parts of the sequence have been investigated, much confusion and misunderstanding have remained. This is because, up to very recently, there had been virtually no regional investigations of the stratigraphy, apart from mapping by the British Geological Survey. This book therefore presents the results of a systematic geological investigation of the Lower Thames region Pleistocene sediments.

The favourable response to my first book *The Pleistocene History of the Middle Thames Valley*, on my investigations into the stratigraphy and palaeoenvironments of the terrace deposits of the Middle Thames and certain of its tributary valleys, encouraged me to write this second, sister volume on the adjacent Lower Thames area sequence in a comparable style. I wrote the first book because I thought it might be appropriate to present all the evidence in a single volume where it could be properly related and assessed. As with the previous study, the work here depends primarily on investigations of the sedimentary sequences and not on terrace morphology. I believe that this synthesis, using as much of the evidence as possible, offers a coherent geological study of the fluvial and related deposits.

As before, the approach I have adopted has been thoroughly interdisciplinary, using evidence from all available sources including archaeology, palaeontology and palaeobotany, as well as the structure and lithology of the sediments themselves. This integrated approach allows the fullest interpretation of the Pleistocene sequence and offers the opportunity of reconstruction of the river's history from the early Pleistocene up to recent times.

In the *Preface* of the Middle Thames book, I remarked that the 'River Thames is perhaps the most widely quoted example of lowland river evolution in Britain [and] I hope that this detailed study of a crucial area will provide both a practical example... [as well as] a stimulus to others'. Since I wrote this there has been a steady increase in the number of studies on British fluvial history. This exciting situation is a welcome development and offers a range of possibilities for future research topics. I very much hope that it continues.

PREFACE

This work represents over eight years of investigations in the Lower Thames region. It was funded by the Natural Environment Research Council and Department of the Environment, to whom I am extremely grateful. Once again there is no doubt that the work could not have appeared in this form if it were not for the encouragement and assistance of many colleagues. I would particularly like to thank Dr M. Aalto, Miss R. Andrew, Dr R. Beck, Dr D. Bridgland, Dr J. A. Catt, Dr G. R. Coope, Mr A. Currant, Mr P. Harding, Dr M. P. Kerney, Dr A. Lister, Dr R. C. Preece, Dr J. E. Robinson, Professor J. Rose, Mr G. de G. Sieveking, Dr C. Turner, Mr J. Whitehead, Mr J. J. Wymer and Dr J. A. Zalasiewicz. Dr C. A. Whiteman assisted me in this research, particularly by doing the clast lithological analysis. To all these people and others who have offered their assistance I would like to express my thanks.

I am as ever indebted to the landowners and contractors who have kindly given me access to their land. The work would not have been possible without the excellent borehole data that were made freely available by the British Geological Survey, the Essex and Kent County Council Highways Departments, the Eastern Road Construction Unit, British Gas, Thames Water Authority and the Greater London Council.

Some of the diagrams were drawn by Justin Jacyno and Mr S. Boreham (Quaternary Research).

I especially wish to thank Professor R. G. West F.R.S. for his unfailing support, encouragement and inspiration throughout this project and indeed my career.

Once again, I must thank Ann for her strength, tolerance, unending support and helping me believe I could finish this work.

This book was written over a period of four years in Cambridge and completed during a period of leave at the Vrije Universiteit, Amsterdam. I thank Professor J. Vandenberghe for his hospitality.

Philip Gibbard
Amsterdam, May 1992

1
Introduction

The Thames Valley is one of the most significant regions in British Pleistocene geology. This is because the valley, together with its extensions into the south Midlands, central Hertfordshire and offshore, contains the longest and potentially the most complete sequence of Pleistocene sediments outside East Anglia. Of crucial additional importance is the position of the valley marginal to the areas that were overridden by glaciers during the cold stages of the Pleistocene. This position enables the Thames deposits to be linked to those of at least two major glaciations. The deposits preserved in the London Basin record a sequence from the marine transgressions of the Pliocene, through initiation and development of the Thames drainage system, to the diversion of the river in the Middle Pleistocene and its subsequent evolution to the present day.

Within the Thames system, the Lower Thames Valley is a classical area since it includes a number of famous fossiliferous and archaeologically important localities in the terrace deposits. For the purposes of this work the area is defined as extending from central London, where the area adjoins the Middle Thames Valley, to the meridian through Stanford-le-Hope, east of Mucking (Fig. 1). This area includes a sequence of terrace surfaces and underlying deposits associated with the present and former courses of the river and its tributaries.

Since initial recognition of the terrace surfaces, a series of investigators have attempted, using various criteria, to separate and identify the different components of the succession. Apart from mapping of the terrace surfaces by the British Geological Survey (BGS), most studies have concentrated on the investigation of individual localities. Until very recently the purely geological aspects of the Pleistocene deposits had been largely neglected, but the success of such studies in neighbouring areas suggested that an investigation of the geology of the Lower Thames Valley using modern methods would yield valuable new observations.

This study represents a comprehensive stratigraphical investigation of the Lower Thames region. By the use of structure, lithology, altitude and stratigraphical position, it has been possible to separate and define deposits of different stages, thereby providing the essential foundation for identifying and correlating units within the region, and also linking the units to the Middle Thames, southeastern and central Essex.

A descriptive section (Chapter 2) contains a definition of each individual unit of the Lower Thames succession, and is followed by detailed discussions of important themes, including pebble lithology (Chapter 3), sedimentary struc-

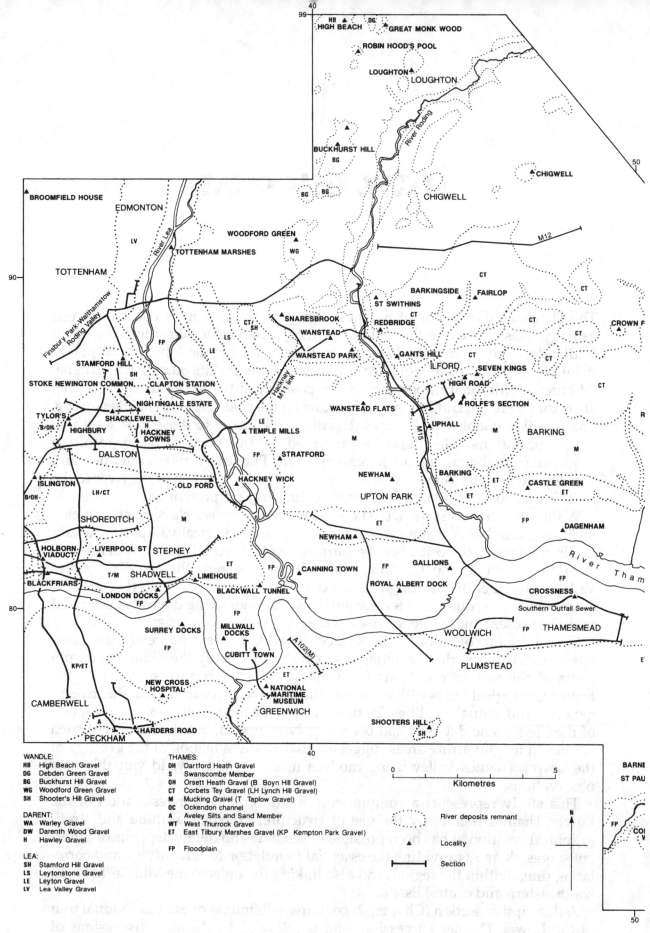

Fig. 1. Location map showing the area studied, the localities of

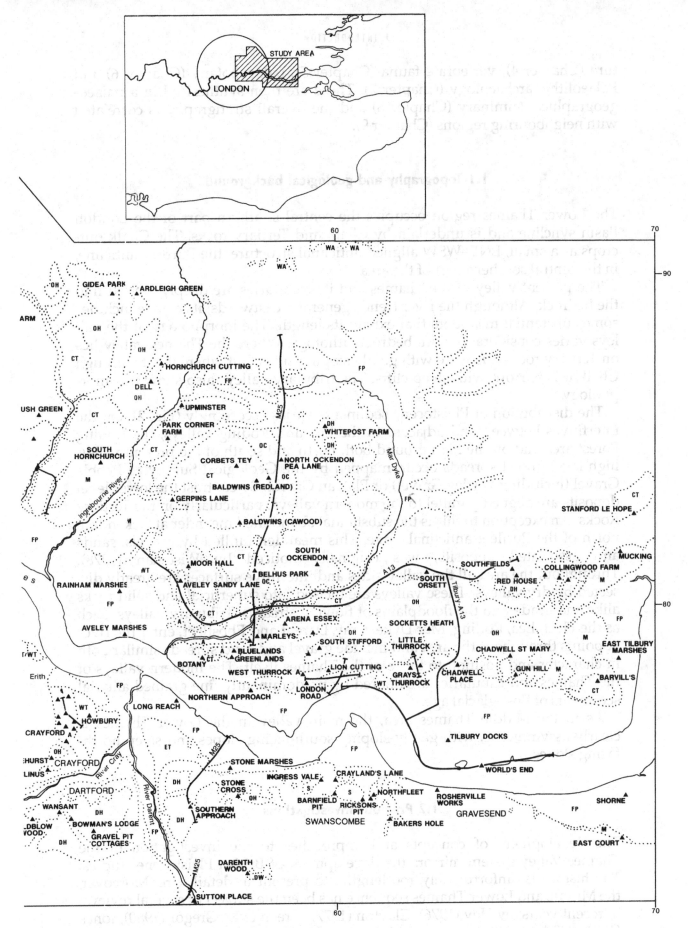

important exposures and borehole sections mentioned in the text.

ture (Chapter 4), vertebrate fauna (Chapter 5), palaeobotany (Chapter 6) and Palaeolithic archaeology (Chapter 7). These are then synthesised in a palaeogeographical summary (Chapter 8) and the overall stratigraphy is correlated with neighbouring regions (Chapter 9).

1.1 Topography and geological background

The Lower Thames region occupies the central southern part of the London Basin syncline and is underlain by Chalk and Tertiary rocks. The Chalk outcrops as a small, ENE–WSW aligned anticlinal structure, the Purfleet anticline, in the central southern part of the area.

The present valley of the Thames and its tributaries are deeply incised into the bedrock. Although the river trends generally eastwards, its course includes some substantial meanders that extend its length. The morphology of the valleys varies considerably with bedrock lithology. Where the Thames valley lies on Tertiary rocks it is wide with gently sloping sides, but where it cuts through Chalk it is narrow, with steep cliffs. The tributary valleys also show this morphology.

The distribution of Pleistocene sediments varies accordingly (Fig. 2). On the interfluves between and adjacent to the Lea and Roding valleys, the Epping Forest area and on the high ground both north and south of the present valley, high-level gravel spreads occur, mapped by the Geological Survey as Pebble Gravel (including Warley Gravel) or Plateau Gravel. By contrast the lower level deposits are aligned parallel to the modern valleys, particularly on the Tertiary Rocks. An exception to this is the substantial abandoned meander at Ockendon, north of the Purfleet anticlinal ridge. This meander, infilled by gravels, sands and associated deposits, was originally noticed by Wiseman (1978). Throughout the area all the valley sides and interfluve surfaces are greatly dissected by dry valleys. These valleys are incised into superficial and solid rocks alike and grade into the floodplains of flowing streams. Tributary valleys, such as the Fleet, Lea, Roding, Mar Dyke, Cray, Darent and Ebbsfleet, enter the river at points throughout the area and include gravel and sand spreads similar both in character and position to those of the Thames itself. The modern valleys of the Thames and tributaries are floored by predominantly fine-grained alluvial sediments of Post-glacial age.

As in the Middle Thames area, the main valley in the Lower Thames is clearly assymetric, with gently sloping south-facing slopes and steep north-facing slopes.

1.2 Previous investigations

The development of concepts and approaches to the investigation of the Thames Valley system mirrors the development of British Pleistocene studies. The history is, unfortunately, too lengthy to present in detail here. Moreover, the Middle and Lower Thames sequence has been the subject of several reviews in recent years, by Hey (1976), Clayton (1977), Green & McGregor (1980), Jones (1981), Gibbard, Whiteman & Bridgland, (1988), Bridgland (1988a) and Gibbard

(1989). It is therefore unnecessary to repeat the reviews in depth here, and instead a short summary of the major contributions is given. The history and usage of individual terrace and unit names are discussed separately at the beginning of each of the unit descriptions in Chapter 2.

During the nineteenth century a number of workers examined the terrace deposits of the Thames Valley. Whitaker (1889) was the first to provide a regional collection of the information on the London region as a whole. He was closely followed by Woodward (1890). They, together with Prestwich (1857, 1890), laid the foundation of the ideas on the evolution of the Thames drainage system. By the turn of the century, Hinton & Kennard (1901, 1907) were able to publish a review of the Lower Thames sediments in which they distinguished four numbered terraces. Following the proposal by Bromehead (1912), that terraces in the Middle Thames should be named from type localities, the terms Boyn Hill, Taplow and Floodplain terraces were subsequently applied during mapping by the Geological Survey to the area east of London. This was published by Dewey & Bromehead (1921), Dewey *et al.* (1924) Bromehead (1925), and Dines & Edmunds (1925). Unfortunately, the geological surveyors mapped Hinton & Kennard's 1st and 2nd terraces together as the Boyn Hill Terrace, a fact that caused some problems for later workers, in spite of the fact that King & Oakley (1936) presented a far more detailed regional review.

Study of the drainage evolution in the London Basin from a geomorphological standpoint by Wooldridge (1927, 1938, 1960) and Wooldridge & Linton (1955) led them to propose a threefold division of the Pleistocene deposits into stages:

I   Summit and high-level deposits; i.e. Red Crag, Pebble Gravels.
II Glacial and Plateau Gravels, including 'Eastern Drift' (Chalky Till).
III Valley and terrace deposits.

During stage I times Wooldridge (1960) concluded that a proto-Thames was aligned through Hertfordshire and Essex. By stage II it had been diverted southward into the so-called 'Middlesex Depression' (Finchley Valley). Subsequent glaciation during his stage II then caused a second diversion of the river and thus initiated the modern Lower Thames Valley. The occurrence of glacial deposits on the northern side of the modern Thames Valley had been known since the mid-nineteenth century. However, the discovery of till (boulder clay or 'Eastern Drift') at its maximum southerly extent at Hornchurch by Holmes (1892) was significant. This was because the till is overlain by gravels and sands of the Boyn Hill Terrace at this site and therefore the relative age of the Lower Thames fluvial deposits to the glacial sediments could be unequivocally established.

The Lower Thames region contains a large number of palaeontologically and archaeologically important localities, several of which have been studied in great detail, but generally in isolation (see reviews in Wymer 1968, 1985). A similar situation previously existed in the adjacent Middle Thames region. However, recent research both north and west of London demonstrated that a detailed sequence could be developed using standard geological techniques.

This work began with the publication by Hey (1965) of his evidence for a 'pre-glacial' Thames flowing from Oxfordshire into Essex based on study of the

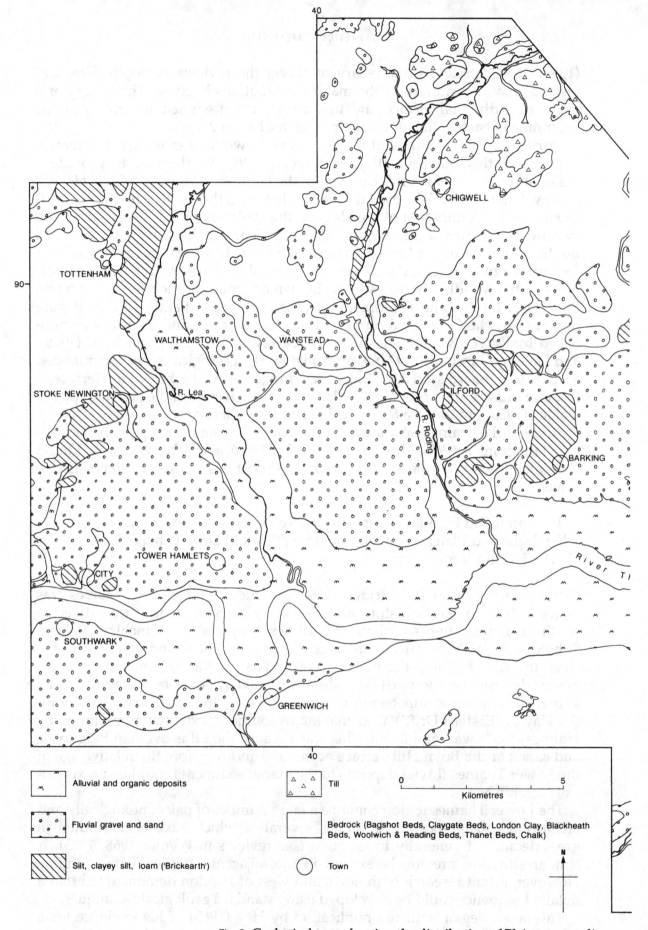

Fig. 2. Geological map showing the distribution of Pleistocene sediments,

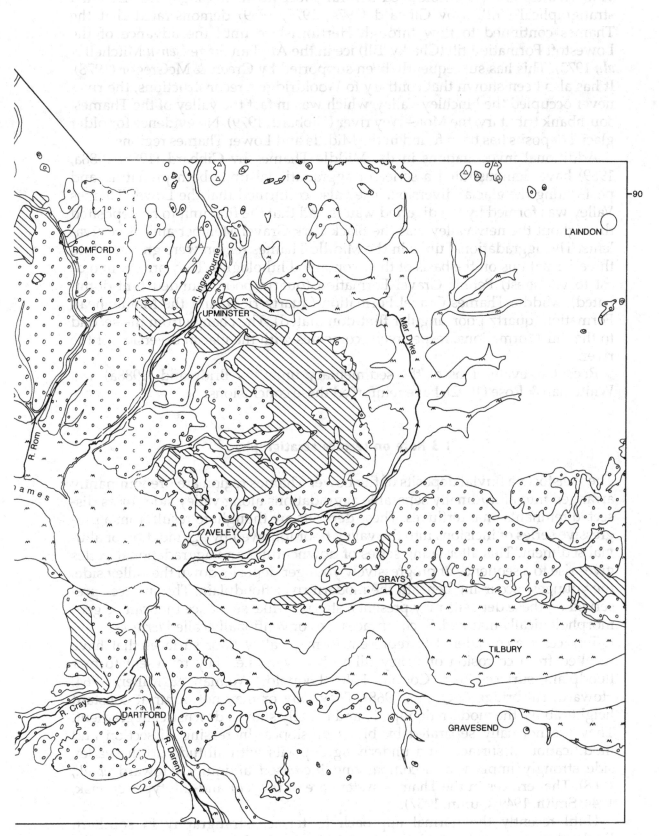

modified from Geological Survey map sheets 256, 257, 270 and 251.

Westland Green Gravels. The success of this attempt to use lithology to subdivide Thames gravels encouraged several subsequent investigations. Detailed stratigraphcal studies by Gibbard (1974, 1977, 1989) demonstrated that the Thames continued to flow through Hertfordshire until the advance of the Lowestoft Formation till (Chalky Till) ice in the Anglian Stage (*sensu* Mitchell *et al.*, 1973). This has subsequently been supported by Green & McGregor (1978). It has also been shown that contrary to Wooldridge's reconstructions, the river never occupied the Finchley Valley, which was in fact the valley of the Thames' southbank tributary, the Mole–Wey river (Gibbard, 1979). No evidence for older glacial deposits has been found in the Middle and Lower Thames regions.

Additional investigations in the Middle Thames by Gibbard (1985, 1988a, 1989) have demonstrated a series of aggradational units both predating and post-dating the glacial diversion. They also confirmed that the Lower Thames Valley was formed by the diverted waters and that the first unit to be deposited throughout the new valley was the Black Park Gravel Member and its equivalents. The aggradational units in the Middle Thames region were grouped into three formations on the basis of their contained lithologies. They are (from oldest to youngest) Pebble Gravel Formation (quartz poor, rounded flint-dominated), Middle Thames Gravel Formation (quartz rich), and the Maidenhead Formation (quartz poor, angular flint-dominated). The change from the second to the third formations, respectively, coincides with the glacial diversion of the river.

Recent re-evaluations of the sedimentary units, by Bridgland (1988a) and Whiteman & Rose (1992), have reinforced these conclusions.

1.3 Note on terrace stratigraphy

The Pleistocene fluvial deposits of the Lower Thames region are predominantly arranged into altitudinally separable aggradations along valley sides or as dissected remnants on interfluves. These deposits are thought to result from a progressive series of incisions into the valley floor followed by aggradation of alluvial sediment. This alternating series of downcutting and aggradational cycles gives rise to a sequence of progressively younger deposits down the valley side, the youngest occurring beneath the modern river floodplain. The surfaces preserved on these deposits may present a stairway-like sequence that can be geomorphologically mapped. Such surfaces are conventionally called *terraces*.

Terraces are considered to represent former valley floor surfaces that have resulted from corrosion or valley filling by rivers, i.e. they represent former floodplain surfaces (e.g. Cotton, 1940; Leopold, Wolman & Miller, 1964; Howard, Fairbridge & Quinn, 1968). They may or may not occur at constant heights above the modern floodplain and usually show a downstream gradient. They are normally separated by bluffs or slopes. Increasing dissection and modification of surfaces and underlying deposits with altitude up the valley side strongly implies an accompanying increased antiquity (Howard *et al.*, 1968). The terraces in the Thames system are of the 'cut-and-fill' type (cf. Fisk, 1944; Smith, 1949; Quinn, 1957).

Until recently the normal approach to terrace stratigraphy in southern England was morphological; that is, subdivision and downstream correlation was based exclusively on altitude. This approach led to the development of morphostratigraphical sequences. However, such an approach, no matter how carefully undertaken, was fraught with difficulties and often ignored the underlying deposits.

In the Thames Valley all exposed terrace surfaces have been modified either by later deposition, soil formation or by periglacial processes. Commonly, valley-side derived surface wash, loess and soliflucted material may blanket the surface, giving rise to the commonly observed concave, rather than horizontal, transverse surface profile sloping down into the river valley. This form is common in the Thames system (cf. Hare, 1947). In addition, degradation of the exposed surfaces by periglacial slope processes, stream erosion, gullying and bedrock solutional subsidence will lead to progressive deterioration in terrace surface preservation with time.

These problems, accompanied by the realisation that the depositional sequences beneath terrace surfaces hold considerable evidence that was previously ignored, has encouraged considerable discussion in the literature concerning the approach to fluvial terrace stratigraphy (for example see Bridgland, 1988b; Gibbard, 1988a). From the Middle Thames investigations (Gibbard, 1985) it was concluded that the most appropriate approach was to identify the lithostratigraphical units preserved beneath terrace surfaces and to define these formally from stratotype localities using standard lithostratigraphical techniques (cf. Hedberg, 1976). Where possible, conventional geological techniques have been used for correlation (avoiding the use of morphostratigraphy). In particular, the thickness of aggradational units is required for downstream or upstream gradient correlation supported by lithology etc. As far as possible, independent age control is offered by biostratigraphical investigation or geochronology where appropriate. In this way an internally consistent and strongly detailed stratigraphical scheme can be constructed. Moreover, this scheme can also be modified and refined as new data become available (see Gibbard, 1985, for further discussion).

1.4 Nomenclature of terrace deposits

Because valleys contain a series of terrace levels, it is necessary to identify them in some way. Early workers used terms such as lower, middle and upper or first, second and third, etc. (Howard *et al.*, 1968). These systems depend on terraces remaining in the same order throughout the valley. As already mentioned, following geological surveying of the Windsor area, Bromehead (1912) commented 'consecutive numbering has frequently been used, but since the 100-feet terrace is spoken alike as the first, second or third ... the interpretation of the literature is attended with much difficulty. For these reasons the adoption of local names for each area is preferable [because] pits [are] wellknown and identifiable, even if future work should reveal higher or intermediate terraces.' Howard (1959) supported this view, concluding that 'numerical systems of terrace nomenclature lack merit and [that] names should be applied as in stratigraphical nomenclature.' The binomial system of nomenclature was subsequently adopted by the Geological Survey following Bromehead's recommendation, and terms such as Taplow Terrace and Boyn Hill Terrace became firmly established.

I INTRODUCTION

In this study a lithostratigraphical subdivision of the sedimentary aggradation has been used throughout, in common with that conventionally established in recent years for the Thames system (e.g. Gibbard, 1985, 1989; Bridgland, 1988a). This subdivision is independent of the terrace surface morphostratigraphy. The latter should ideally be viewed as being developed *upon* the depositional sequence beneath. As such, a parallel nomenclature may be required for morphostratigraphical purposes if this is thought to be necessary in the future. However, no such terminology is offered here.

2
Lithostratigraphy

The Pleistocene deposits of the Lower Thames Valley have been studied in detail from exposures and borehole sections, the location of which are shown in Fig. 1. The deposits have been subdivided following the International Stratigraphic Guide (Hedberg, 1976) in which aggradations are defined and delimited as far as possible as formal units from stratotype localities. These units are considered to be of member status because this is the most appropriate hierarchical level and is compatible with neighbouring areas, particularly the Middle Thames Valley (Gibbard, 1985, 1988a, 1989). This is contrary to the proposal by Bridgland (1988a, b) that aggradations should be assigned formation rather than member status, because the former is the 'primary unit (in lithostratigraphy)'. Whether this argument is theoretically correct or not is not in dispute. The real problem stems from the concepts of compatibility and practicality (Gibbard, 1988a). The present author considers the formation to be a large-scale unit, as applied in neighbouring areas (e.g. Kesgrave Formation, Lowestoft Formation, North Sea Drift Formation, Red Crag Formation), and therefore in this text the term formation is retained to refer to a collection of formally defined members with broadly unified characteristics, as in earlier publications.

A similar argument applies to the next unit in the lithostratigraphical hierarchy, the Group. This term is considered too large scale for use in Pleistocene stratigraphy, by comparison with Group status units in other parts of the geological column. No units of this rank will be used or defined here.

In order to obtain as detailed a sequence as possible the deposits have been investigated using interdisciplinary methods. In the field a careful examination of the entire area was undertaken to locate and study exposures. All exposures were cleaned and logged, with significant features being recorded. All sections were levelled to Ordnance Datum. Palaeocurrent measurements were undertaken wherever possible and fabric determinations made whenever appropriate. When required, sections were sampled for pebble lithological counts and grain-size analysis. Where fossiliferous sediments were encountered they were also sampled either in section or by borehole. Duplicate samples were then subdivided for various palaeontological analyses. If possible, samples for absolute radiometric and thermoluminescence techniques were also collected. Where no exposures were available holes were dug or augered and larger diameter boreholes were put down to investigate the nature of the sediments.

The methods for individual general techniques used were as follows:

1. Pebble lithological counts were made on samples of pebbles in the 33–8 (–5 to –3ϕ) range. All samples were washed and sieved to remove superfluous material. Pebbles were then sorted into lithological groups, clasts of questionable lithology being broken. The numbers of pebbles in each group were then counted and the percentages of each of the component lithologies determined. The groups identified follow those mentioned by Gibbard (1985), together with some types from Bridgland (1986). When sampling, care was taken to avoid the zone of periglacial disturbance immediately beneath the surface, in order to obtain unmodified gravel.
2. Palaeocurrent analyses were measured on foreset beds in sand and pebbly sand following the recommendations of Potter & Pettijohn (1963). The results are expressed as vector means.
3. Pebble long-axis fabrics were performed by measuring the dip and dip-direction of the long axis of elongated clasts from a small area of vertical face. The measurements were then plotted on a Schmidt net and the resulting eigenvectors calculated using the Stereo computer program.
4. Grain-size analyses were undertaken using the hydrometer and sieve techniques following the ASTM (1964) standard procedure using 50 g of air-dried sediment matrix of less than 2 mm (–1ϕ) diameter. Results are shown as cumulative percentage curves.
5. Borehole data were collected from various sources and interpreted where necessary by the present author. Wherever possible, suspect records were checked either by comparison with neighbouring sections, or other records.

The borehole data have been assembled from several sources. In the descriptive sections the source of individual borehole logs is not stated, for the sake of brevity. However, borehole data have been obtained from the British Geological Survey (BGS) borehole records library, unless otherwise stated. Other sources include Essex County Council Highways Department (ECC), the Greater London Council (GLC), Kent County Council Highways Department (KCC) and the Eastern Road Construction Unit (ERCU). The records for all the holes are also available from the author.

Sections drawn from these borehole records have been interpreted in the light of field experience and checking by the author. The localities of sections illustrated are shown in Fig. 1. The figures illustrating the sections use the sediment symbols shown in Fig. 3.

In addition to the borehole sections, long-profile summary geological sections have been constructed for each of the major aggradational units defined. These sections are based on all the data sources available and discussed in the text.

The units defined using these methods are described below in stratigraphical order and their distribution is shown in Fig. 1. Tributary valley equivalents are discussed immediately following the appropriate Thames member to which they relate.

2.1 High-level ('pre-Thames') deposits

Both north and south of the present Lower Thames Valley, highly fragmented spreads of deposits occur at heights considerably above those reached by local

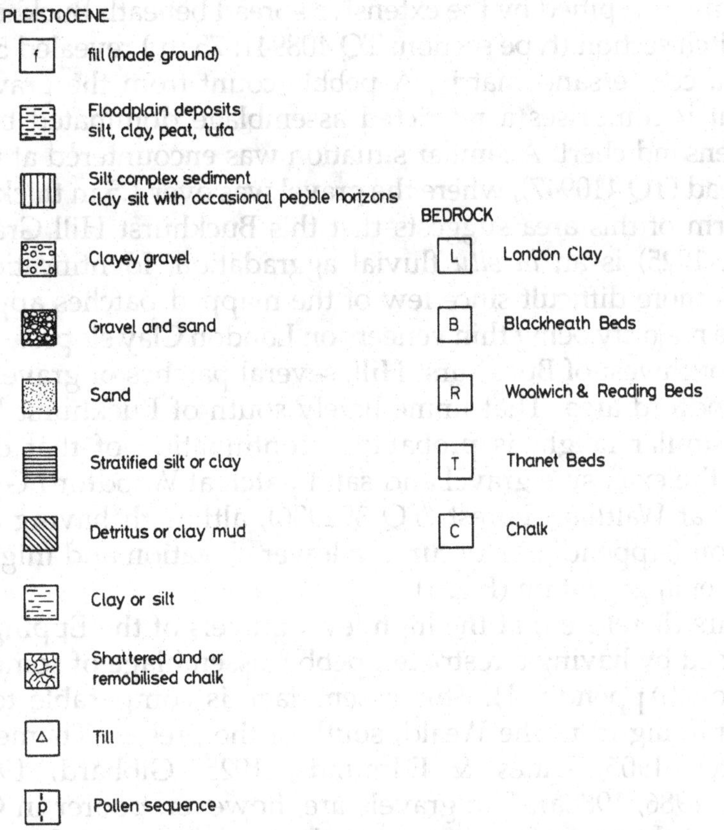

2.1 HIGH LEVEL DEPOSITS

Fig. 3. Standard sediment symbols used in all geological sections in this work.

Thames sediments. They comprise two groups: gravels and sands, and glacial gravels and till, which are described separately below.

(a) High-level gravels

The best development of pre-Thames gravels in the area occurs on the Epping Forest interfluve between the Rivers Lea and Roding (Fig. 2). They are also present near Chigwell. On the geological maps (sheets 256, 257), these gravels are shown as 'Pebble Gravel' and 'Glacial Gravel'. The former have not been investigated during this study, but are described by Whitaker (1889), Bromehead (1925) and Dines & Edmunds (1925). The gravel with a coarse sand matrix caps the hilltops as isolated outliers. It is usually 1–1.5 m to rarely 2 m thick and comprises predominantly flint with subordinate quartz and Greensand chert pebbles. No satisfactory exposures of these sediments were available during this study, but examination of surface and ditch scatters confirm earlier observations. However, Pebble Gravel occurs at two distinct levels, that at over 100 m at High Beach (TQ 408978: 109 m) and that at over 90 m at Debden Green (TQ 429981: 93 m). A pebble count from the latter illustrates the restricted assemblage (Appendix 1).

Investigation of the lower level patches, mapped as glacial gravel, reveals a more complex situation. Those recorded in the northern part of the area include some yielding a limited pebble assemblage, closely comparable to that of the local Pebble Gravel, but others contain a more diverse assemblage similar to the glacial gravel of the area (see below).

The former is typified by the extensive spread beneath Buckhurst Hill, where a shallow ditch section (type section: TQ 408941: 75 m) revealed 55 cm of medium gravel in a coarse sand matrix. A pebble count from the gravel (Appendix 1) shows that it comprises a restricted assemblage dominated by flint, but with some Greensand chert. A similar situation was encountered at the northern end of the spread (TQ 410947), where the gravel was over 1.5 m thick. The terrace-like surface form of this area suggests that this Buckhurst Hill Gravel (cf. Dines & Edmunds, 1925) is an *in situ* fluvial aggradation. Identification of equivalent deposits is more difficult since few of the mapped patches appear to be undisturbed, the majority being thin veneers on London Clay slopes.

South-southwest of Buckhurst Hill, several patches of gravel and sand occur in the Woodford area. That immediately south of Buckhurst Hill (TQ 402926) and at a similar height is probably a continuation of that described above. However, the extensive gravel and sand patch at Woodford Green (TQ 395911: 67 m) and at Waltham Forest (TQ 392906), although having a similar pebble composition (Appendix 1), occur at a lower elevation and might therefore represent a later aggradation (Fig. 4).

It appears therefore that the high level gravels of the Epping Forest area are characterised by having a restricted pebble assemblage of somewhat consistent composition (Appendix 1). This assemblage is comparable to that typical of streams draining from the Weald, south of the present Thames (cf. Prestwich, 1890; Salter, 1905; Dines & Edmunds, 1925; Gibbard, 1979, 1982, 1985; Bridgland, 1986, 1988a). The gravels are, however, poorer in Greensand chert than those of the Mole–Wey stream to the west. It would therefore seem possible to identify a distinct assemblage for the deposits described here and so they are assigned to the Epping Forest Formation.

On the basis that the deposits are of Weald origin, upstream projection of the valley trend (towards SSW) strongly suggests that the stream responsible was the Wandle. The fluvial sequence in the Wandle valley has been extensively studied by Peake (1971, 1982) who recognised two high level terrace remnants; a Norwood (Crystal Palace)Terrace at 112 m OD and an Effra Terrace at 92 m OD in the Norwood area. The former, originally investigated by Prestwich (1890) and more recently by Macklin (1981), is underlain by braided stream deposits, containing *c.* 10% Greensand chert, derived by the river from a gap at Caterham or Mertsham (Dewey & Bromehead, 1921). The lower, younger unit may be of similar origin. As can be seen in Fig. 4, the 31 km from Woodford to Crystal Palace is probably too far for reliable correlation by gradient (the gradient required being 1.4–1.6 m km^{-1}). Nevertheless, if the gradient is of the correct order, it is possible that the Buckhurst Hill or Woodford Green Gravel could be equivalent to the Effra Gravel and the Debden Green Gravel, the Norwood Gravel. The low frequency of Greensand chert in the Epping Forest Formation gravels compared with those at Norwood must have arisen by downstream incorporation of flint from local rocks, e.g. the Claygate Beds, thus diluting any Weald-derived components.

High-level gravels also occur south of the Thames at Shooter's Hill and in Darenth Wood. At the former a gravel patch caps the hill at 130 m OD (TQ 436765). The Shooter's Hill Gravel (Dewey *et al.*, 1924) has been repeatedly studied by Leach (1912, 1920) who recorded a section at Shrewsbury Lane Fire Station where it comprised 3.7 m of interbedded gravel and coarse sand resting

2.1 HIGH LEVEL DEPOSITS

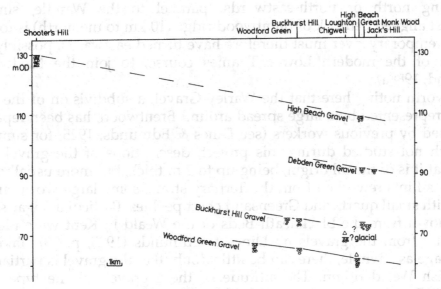

Fig. 4. Long profile section of the high-level gravel spreads in the Epping Forest area, showing correlations described in the text.

on London Clay. The deposit is generally described as being 'red-stained' – a reference to strong iron oxide deposition typically found in gravels of considerable antiquity. The pebble assemblage comprises 'Eocene flint pebbles, a few more or less worn flints from the Chalk and rarely little fragments of Lower Greensand chert and small quartz pebbles' (Dewey *et al.*, 1924 p 84). No exposure was available during the present project; however, the consistency and detail of the descriptions imply that they are reliable.

The similarity of the lithological assemblage to those discussed above clearly indicates that the Shooter's Hill Gravel must also have been deposited by a river draining the Weald. The identity of the stream responsible is difficult to determine. However, there are only two possibilities: either the Darent–Cray system or the Wandle. The position of the deposit on Tertiary rather than on Chalk bedrock and its close proximity to the SSW–NNE course proposed for the Wandle possibly imply that the gravel represents an early aggradational remnant of this stream. This possibility is strengthened by downstream altitudinal correlation that suggests possible equivalence to the High Beach Gravel (Pebble Gravel), since gravel of comparable lithology is absent from the hilltops between Havering and the Roding Valley (cf. Dines & Edmunds, 1925).

At Darenth Wood, south of Swanscombe, a patch of 'Plateau Gravel' is shown on the geological map (sheet 271) to be resting on Woolwich Beds. This gravel is very poorly exposed, but on the north side of a footpath (TQ 572718: 80 m) a shallow deposit (20 cm thick) occurred, comprising abundant Tertiary-type rounded flint, with brown angular flints, rare small quartz and Lower Greensand clasts (a satisfactory sample could not be obtained). This Darenth Wood Gravel, originally recorded by Prestwich (1890) and more recently by Dewey *et al.* (1924) is again of Weald origin, presumably having been deposited by the Darent. No downstream correlatives are known, and therefore the contemporary course of the river cannot be determined. Nevertheless, downstream extension of a gradient of 3 m km^{-1} (that of the modern Darent), or as shallow as 1.5 m km^{-1}, clearly indicates that the stream would have been prevented from continuing north or north-eastwards, parallel to the Wandle, since the east–west aligned bedrock of Brentwood ridge (10 km to the north) is too high. The contemporary river must therefore have turned eastwards, possibly along the line of the modern Lower Thames course, to join the Medway (cf. Bridgland, 1988a).

It is worth noting here that the Warley Gravel, a subdivision of the Pebble Gravel, represented by a large spread around Brentwood, has been repeatedly recognised by previous workers (see Dines & Edmunds, 1925, for summary). Although not studied during this project, descriptions of the gravel clearly show that it is of fluvial origin, being up to 2 m thick, but more usually 1 m. It comprises flints reworked from the Tertiary strata, some large worn (angular) flints, with small quartz and Greensand chert pebbles. Particular 'cigar-shaped' flints known from the Blackheath Beds of the Weald in Kent were identified commonly from the gravels by Dines & Edmunds (1925, p. 23). Taking the assemblage as a whole, there can be little doubt that the gravel is partially also of Kentish Weald origin. The altitude of these gravels at the type section (TQ 513915: 109 m) suggests a considerable antiquity, compared with those already described. Similar material was discovered at Navestock Side (TQ 563975: 96 m OD). They may therefore record a very early course of the Darent–Cray north of the Thames Valley, as suggested by Wooldridge (1927). Clearly, therefore, the diversion of the Darent from this northerly alignment to the east to meet the Medway must have occurred during the interval between deposition of the Warley Gravel and that of the Darenth Wood Gravel.

(b) Greenhithe chalk quarry

In 1983, the south face of the Greenhithe chalk quarry (TQ 589744) exposed a large, vertically orientated, pipe-like solution hollow. This feature extended through almost the entire vertical height of the face (c. 9.5 m) from beneath an impersistent bed of reddish-brown to yellow silt sand to pebbly clay up to 2 m thick. It penetrated both the 1.25 m of Thanet Sand and at least the 8 m of chalk bedrock exposed beneath. The hollow reached a maximum width of 3.5 m and was infilled by vertically orientated beds of sorted gravel and sand. The long axes of the pebbles were also orientated vertically, suggesting that the material had collapsed or been washed into the hollow. The pebble count (Appendix 2) comprises only material of local and Weald origin. The position of this isolated material, its lithology and sedimentary structures suggest a fluvial origin. The most probable origin in this area is as a deposit of the River Darent.

(c) Glacial gravel and till

Chalky till is shown on the Romford geological map (sheet 257) over much of the northern part of the sheet. The highly dissected till impinges on the area studied in two main places: the Roding Valley as far south as Loughton and Chigwell, and in the Noak Hill–Hornchurch area. The till sheet is generally regarded as being the southernmost extension of the Lowestoft Till Formation, which underlies much of southern East Anglia (Dines & Edmunds, 1925; Perrin, Rose & Davis, 1979; Whiteman, 1987; Allen, Cheshire & Whiteman, 1991). Associated with the till are sorted sands and gravels, the lithology of which includes exotic elements that can only have been brought into the area by a glacier (chapter 3).

2.1 HIGH LEVEL DEPOSITS

The Lowestoft Till typically has a blue-grey to brown matrix and is rich in chalk and flint clasts, as well as containing a variety of materials derived from the Jurassic and older rocks of central to northern Britain (Perrin *et al.*, 1979). It was deposited by ice moving broadly southwards in this area (Allen *et al.*, 1990). Although not specifically studied during this project, mention must be made here of the till recorded originally by Holmes (1892, 1894) at Hornchurch. Here sections cut for a railway line from Romford to Upminster exposed chalky till up to 4.5 m thick, resting directly on London Clay and overlain by fluvial gravels and sands, 3–4 m thick. Recent re-excavation of a section in the cutting at TQ 54708737 for English Nature by Bridgland, Cheshire, Gibbard, Harding & Whiteman (unpublished) confirmed and improved Holme's original descriptions (Fig. 5). For example, fabric measurements from 15 cm above the till base are strongly aligned almost east–west (Fig. 5). At first sight this appears contrary to the known regional ice movement and till distribution, but the fabric can be interpreted as transverse. Such fabrics are known to develop where compressional flow occurs, such as in front of an obstacle. If this is the case at Hornchurch, it would indicate deposition by ice moving from the north-east.

Whiteman (1987) has correlated the Hornchurch Till with the Newney Green Member of the Lowestoft Till Formation in Essex.

The height of the till base at Hornchurch (*c.* 32 m OD) and its position, directly overlying London Clay bedrock, suggest it was emplaced in a trough, possibly an overdeepened valley, eroded by the ice. The lack of fluvial deposits beneath the till anywhere in this trough indicates that the existence of a pre-

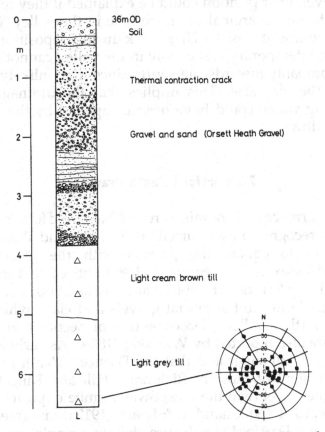

Fig. 5. Section excavated by the Nature Conservancy Council in the Hornchurch railway cutting and clast fabric of the Hornchurch Till.

glacial stream valley here is very unlikely. This suggestion is supported by the occurrence of the till at a somewhat lower level than the projected altitude of the first Thames' member through the Lower Thames Valley (see below).

East of Loughton, two gravel patches occur adjacent to the Roding Valley at approximately comparable altitudes to that at Buckhurst Hill. However, the 70 cm thick gravel at Robin Hood's Pool (TQ 415968: 74 m) contains a slightly more diverse pebble assemblage than at Buckhurst Hill (Appendix 1). Likewise, the 60 cm thick patch, beneath Great Monk Wood (TQ 423976: 73 m) also includes exotic lithologies, notably *Rhaxella* chert. It is unlikely therefore that these two patches are equivalent to the Buckhurst Hill spread, but, as suggested by Dines & Edmunds (1925), they are probably of glacial origin (see below). The same authors record that gravel and sand also underlie till at Loughton (TQ 430963: 62 m) and Chigwell (TQ 467932: 69 m) at a lower height. At neither locality could satisfactory samples be obtained, but at the latter a borehole demonstrated 45 cm of silty sand containing occasional flint pebbles, beneath decalcified till (cf. Whitaker, 1889). Moreover, the former could represent a downstream equivalent of the gravel at Woodford. If so, then, by analogy with the Finchley and Watford areas (cf. Gibbard, 1977, 1979), the occurrence below *in situ* till of stratified sediment in an apparently undisturbed position suggests that it might have been the final unit deposited by the stream before the Lowestoft Till ice lobe advanced into the valley.

If this interpretation is correct, then the Great Monk Wood–Robin Hood's Pool deposits are apparently too high to be part of the same unit as that at Loughton. However, their position could be explained if they represent a kame terrace remnant formed marginal to an ice lobe in the valley. This conclusion implies that the ice was at least 15–20 m thick during deposition of the feature. The direction of contemporary water flow in the valley cannot be determined, but here it was probably towards the south, since it is unlikely that the water flowed towards the ice sheet. This implies that the drainage reversal that formed the Roding valley could have been completed by the time that these gravels were laid down.

2.2 Dartford Heath Gravel

The high-level occurrence and contained rolled Mesozoic fossils of the Dartford gravels were first recognised by Spurrell (1880, 1886) and Wood (1882). These features led them to equate the gravels with the glacial deposits of Hertfordshire and Essex. Indeed, Spurrell (1886) noticed that the gravels were the highest containing 'erratics or pebbles and fossils of northern origin, which have been washed from the the glacial gravels and clays which still line the northern brow of [the] valley'. Observations of sections in the gravel on Dartford Heath are also reported by Whitaker (1889). As early as 1905, Hinton & Kennard (1905, p. 80) suggested that the Dartford Heath gravels were the downstream equivalent of those at Richmond Hill and Kingston, since they occurred at a higher level than the local Swanscombe deposits. This was supported by Woodward (1909). Chandler & Leach (1912a) disagreed and grouped the Swanscombe and Dartford Heath spreads into a single aggradation on the basis of sections at the Wansant Pit. The Geological Survey mapping (Dewey *et al.*, 1924) followed this interpretation, the gravel spread beneath Dartford Heath being included with the Boyn Hill Terrace, in which the Swanscombe sections were also thought to occur. However, Zeuner (1959, p. 154–5) reasserted Hinton & Kennard's (1905) views and concluded that the downstream gradient indicated that the Dartford Heath Gravel was indeed an earlier event that predated the local Boyn Hill Terrace deposits. However, not all workers agreed with this view. For example, Wymer (1968) still grouped the Dartford Heath and Swanscombe deposits together. Despite this, recent research has reinforced Zeuner's conclusions. Gibbard (1979, 1985, 1989) agreed that the gravels underlying Dartford Heath were the highest and therefore the oldest Thames deposits present in the area. These he equated with the late Anglian Black Park Gravel upstream. This correlation means that the Black Park/Dartford Heath Gravels provide an important horizon of known age throughout the Thames system.

The deposits underlying Dartford Heath were well exposed in the late 1970s to early 80s in the southernmost part of the Wansant Pit (type section: TQ 514737: 42 m) where they are 4.25 m thick (Gibbard, 1979). They comprise horizontally bedded medium gravel with a sand matrix, interbedded with horizontally bedded sand units 25–35 cm thick (pebble count: Appendix 2). Repeated examination of remaining sections at this site show the gravels rest on a gently undulating surface of the Thanet Sands. Nowhere is there evidence for the continuity of aggradation with the lower level Swanscombe deposits, reported by Chandler & Leach (1912a), and therefore their observation cannot have been correct. However, it is clear from descriptions by Wymer (1968, pp. 326–7) that sections may have been more complex in the past. Repeated mention of 'loam-filled channels' or a covering of the surface of the gravel (e.g. Smith & Dewey, 1913) clearly indicates a younger accumulation than that beneath, possibly related to the sediments in the neighbouring Bowman's Lodge Pit. According to Wymer (1968) only two ovate hand axes are known from the gravels themselves.

Immediately east of the type section is the Bowman's Lodge site (TQ 518738). Although infilled at the time of the present study, this site has previously yielded an important assemblage of unrolled Palaeolithic artefacts. According to Tester (1951), the industry occurred at the contact of the underlying gravel and an overlying brown loam bed. The latter varied in thickness from 0.3–4.6 m and was thought to be the lateral equivalent of that capping the sections at the Wansant site. Detailed examination of the area immediately south-east of the Bowman's Lodge Pit by auger indicates that a brown clayey silt-filled channel up to 2 m deep is present. This channel follows an arcuate course from the infilled pit across the Heath to immediately north of Leyton Cross and continues into the dry valley that extends east-northeastwards to the Darent Valley, south of Dartford town centre (Fig. 1). There can be little doubt that this channel post-dates the deposition of the Dartford Heath Gravel and is related to the formation of a local drainage system on the terrace surface.

Other gravel sections on Dartford Heath were in a poor condition at the time of investigation. However, old pits close to the B2210 Brooklands Road (TQ 521734: 40 m) exposed 4.5 m of stratified gravel similar to that at the type site resting on Thanet Sand. The upper 20 cm, immediately below the ground surface, was cryoturbated. A comparable exposure was present at Gravel Pit

Cottages (TQ 525731: 39.5 m). South of the A2 road, the deposits were visible in Coldblow Wood (TQ 508737: 32 m), but attempts to find other sections in the vicinity failed. However, sections were seen here by Dines et al. (1924, p. 96), who noted the absence of northerly erratics and therefore suggested that the gravels represent the contemporary confluence of the Cray, Darent and Thames. This is supported by the pebble assemblage obtained here (Appendix 2).

On the basis of gradient projection and field survey, much of the country in the Dartford area mapped as Boyn Hill Terrace falls into this member: for example, the gravels and sands exposed in the extensive cuttings for the Dartford Tunnel southern approach road (M25; Fig. 6). Here sections extended from 556741 to 557736 (39–42 m OD). The deposits were up to 6 m thick and again comprised predominantly horizontally bedded medium gravel, individual beds being 30–40 cm thick, and sand, either as channel-like lenses or as more continuous beds up to 20 cm thick. The deposits rested directly on an undulating Chalk surface, which was much affected by solution. The latter appeared as narrow vertical pipes, up to 1 m in diameter and became markedly more intense towards the northern end of the sections where the deposits thinned. Here the cryoturbation of the upper part of the gravels, mentioned above, intersected with the solution features to form a highly contorted sediment complex. In addition, boreholes at 736557, immediately west of the road, have proved an anomalous depth of over 10 m gravel and sand, overlain by 8.7 m of sand and in turn capped by 9.35 m of yellow-brown clayey silt. The limited extent and form of this filling suggests a doline-type feature. The Dartford Heath Gravel is cut out to the east by later dissection, associated with the lower Orsett Heath Member, the latter represented at Stone Cross (TQ 567742: 31 m).

Downstream, the deposits of this unit are not again encountered on the south side of the river. However, on the basis of downstream gradient, the spread of gravels, mapped in Essex as Boyn Hill by the British Geological Survey around the Orsett Cock public house, can be equated with this member. A small pit at Southfields (TQ 656812: 36 m) exposed a total thickness of 4 m of deposits, the lower 1.5 m of which comprised large-scale tabular cross-stratified sands with pebbly cross laminae. This was overlain by stratified medium to fine gravel (pebble count, Appendix 2), giving way to cross-bedded gravel and interbedded sand in the upper metre. The uppermost 1 m was cryoturbated and locally rather reddish in colour. Palaeocurrent measurements indicate a flow towards ENE. The Pleistocene sediments at this site rested directly on a pebble bed in the Blackheath Beds. The gravels are also encountered in boreholes for the A13 road (Fig. 7). Sections in this area have also been common in the past, borehole data (ECC) for the extensive Red House Pit (TQ 656809: 37 m) proved 4 m of gravel and sand. Similarly, at Collingwood Farm (TQ 664808: 34 m) 5 m of tabular cross-bedded sand and fine to medium gravel occurred, giving a palaeocurrent direction towards the east. In the west face the section is capped by 1.25 m of light-brown clayey silt with occasional pebbles in the basal 30 cm (section 2.10).

Immediately to the east, further sections south of St Clere's Cottages expose similar deposits resting on an undulating or channelled surface of Thanet Sand at TQ 670810, 809669, 806669 and 808672 (33 m OD). The maximum thickness here was 5.5 m and comprised 10–15 cm thick tabular cross-bedded sand with thin pebble interlaminae, giving way to fine to medium gravel and again capped by 1–1.25 m of brown silt or clay.

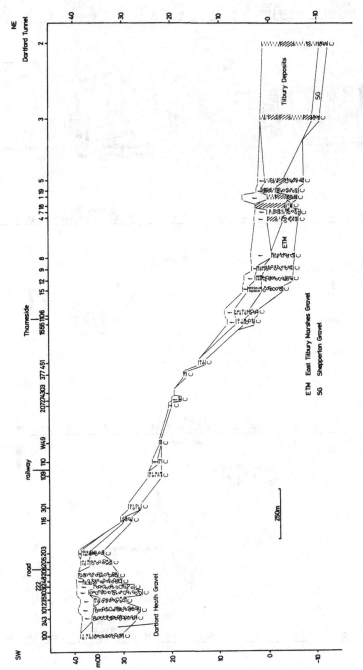

Fig. 6. Section along the M25 motorway southern approach to the Dartford Tunnel, constructed using borehole data (source: KCC).

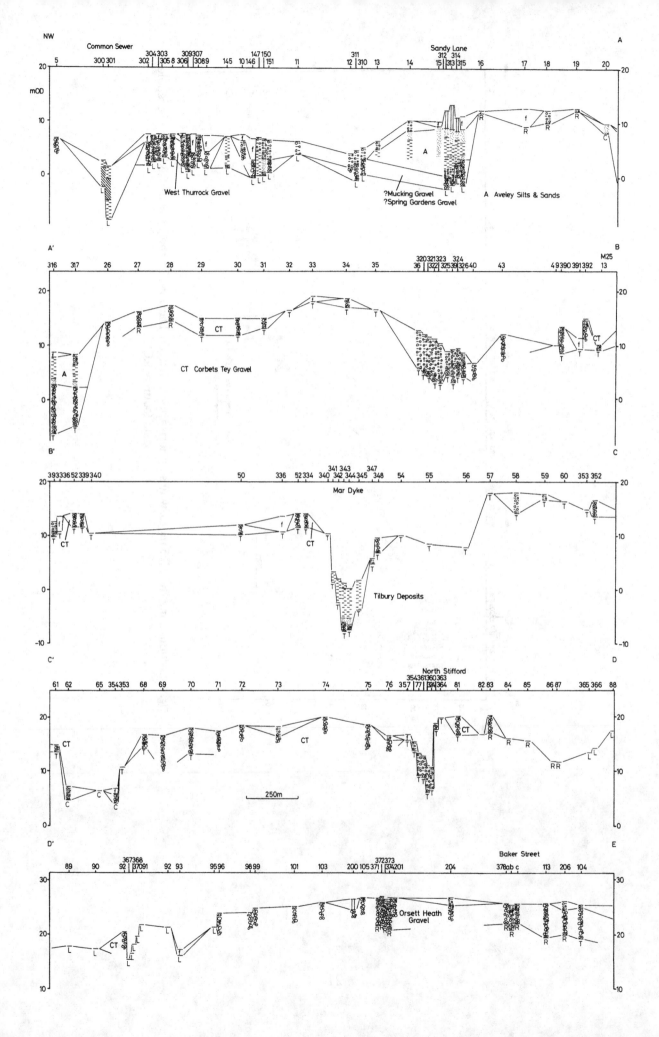

2.2 DARTFORD HEATH GRAVEL

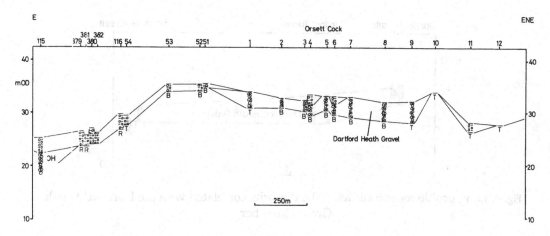

Fig. 7. Section along the realigned A13 road from Rainham to Orsett Cock, constructed using borehole data (sources: ERCU, ECC).

Upstream in the Ockendon area, a small outlier of the unit mapped east of Whitepost Farm (TQ 596854: 41 m) by Dines & Edmunds (1925) could not be confirmed during this project but has been identified in recent remapping by the BGS. To judge from its altitude and position it could represent an upstream equivalent of this unit.

Pebble assemblages from the Dartford Heath Gravel are characteristic because they are the highest and therefore the oldest gravels in the Lower Thames valley that contain northern constituents (see above). They include 39–69% angular flint, 21–51% rounded ('Tertiary') flint, 3–8.5% vein quartz and quartzite and 2–6.5% Greensand chert (Appendix 2). Down-valley and vertical sequence changes occur. Particularly striking is the downstream enrichment in rounded 'Tertiary' type flint derived from local pebble beds. Also a locally high frequency of Greensand chert accompanied by high rounded flint counts occur at some of the Dartford sites (e.g. the Tunnel approach), indicating possible confluence of the Darent stream in this area. Of importance is the recognition of the exotic *Rhaxella* chert from these gravels (Smith & Dewey, 1913, p. 199), which is thought to have entered the Thames' system from the Anglian glacial ice (Bridgland, 1983a). Other northern erratics of similar derivation have also been recovered.

The resulting gradient for the Dartford Heath Gravel, obtained by plotting the localities mentioned, is 50 cm km^{-1} (Fig. 8). This corresponds closely with that for upstream correlative the Black Park Gravel (Hare, 1947; Gibbard, 1979, 1985). This evidence reinforces previous views that the Dartford Heath Member is the first unit deposited by the Thames aligned through the Lower Thames valley after the river's diversion.

The uniform lithology of mainstream Thames deposits in the Lower Thames valley from the Dartford Heath Gravel to the youngest deposits implies that this and all younger members should be grouped into a single Lower Thames Gravel Formation.

The brown silt overlying the gravels at some sites appears to represent a slope wash or mass flow deposit possibly derived from London Clay, to judge from the indistinct stratification and variable thickness. The high silt content may include a loess component (section 2.10). The red staining, recorded at Southfields, may be of pedogenic origin.

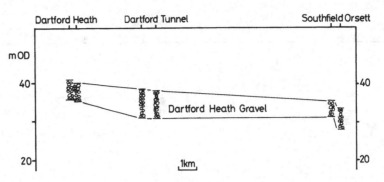

Fig. 8. Long profile section showing the deposits correlated with the Dartford Heath Gravel Member.

2.3 Swanscombe Member Deposits

The Pleistocene deposits at Swanscombe are the most intensively studied of any in the Thames Valley. This is because more Palaeolithic artefacts have been found here than at any other site in Britain. Many of the artefacts are in unrolled condition and are associated with abundant finds of fossil vertebrates and Mollusca (Wymer, 1968).

Although finds were made in the area from the 1880s onwards, systematic investigations did not begin until 1912 when Smith & Dewey first established the sedimentary sequence and its contained artefact assemblages at Barnfield and Colyer's Pits (these two sites were later combined and are collectively referred to as Barnfield Pit). They continued their work the following year (Smith & Dewey, 1913, 1914) at the neighbouring sites: Ingress Vale (Tavern or Dierden's), Crayland's Lane and the lower Wansant Pit (Fig. 1). Although considerable finds continued to be made, controlled excavation did not take place again until 1937 following the discovery of human skull fragments by Marston (1937). Later excavations were carried out in 1955–60 (Wymer, in Ovey, 1964) and 1968–72 (Waechter, 1969, 1970, 1971, 1973). Additional reports on the sequence have been published by Oakley (1952), Ovey (1964), Wymer (1968), Kerney (1971), Evans (1971), Waechter & Conway (1977), Bridgland et al. (1985) and Kemp (1985). On the basis of the fossil assemblages, the sequence is thought to span the Hoxnian temperate Stage.

Since so much information is already published on the Swanscombe sequence, and because sections were only available during this study at Barnfield (in 1982, 1987) and Crayland's Lane Pits, the stratigraphy summarised below is based mostly on earlier reports, particularly Bridgland et al. (1985). Some minor interpretative differences between earlier writers' views and those of the present author are included.

The information here concentrates on three major sites: Ingress Vale (Tavern or Dierden's: TQ 595748), Barnfield (including Colyer's: TQ 595745) and Rickson's Pits (TQ 611743), but Smith & Dewey's (1913, 1914) sequence, with modifications by Marston (1937), provides one that is still recognised today. The sequence is illustated in Fig. 9.

2.3 SWANSCOMBE MEMBER DEPOSITS

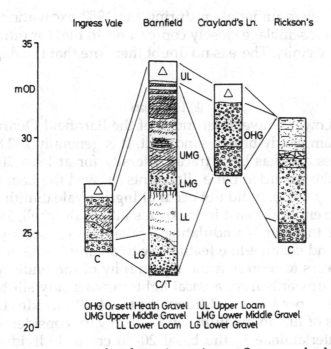

Fig. 9. Sediment sequences exposed at the main sections at Swanscombe, based on sources quoted in the text.

(a) Lower Gravel

The basal deposits rest on a gently undulating bedrock surface at 22.75–23 m OD at Barnfield Pit. This surface exposes Thanet Sand, but this thins northwards across the site to expose the Chalk beneath. Both Bridgland et al. (1985) and Conway (1973) recognise a division of the Lower Gravel into two subunits: the lower subunit comprises clast-supported, medium gravel in a sand matrix, 1.2 m thick. Large flint cobbles are present at the base and a lens of tabular cross-bedded pebbly sand, 50 cm thick, occurred above. Shells are absent in the basal part and present only as broken fragments above. According to Conway (1973), vertebrate remains recovered from this subunit included a fallow deer *Dama dama* antler, a horse *Equus* tooth and an aurochs *Bos* metacarpel.

The upper subunit is also 1.2 m thick and consists of finer gravel with an increased sand content, both as matrix and interbedded sand laminae. The basal lamina of this subunit contained abundant chalk clasts, according to Conway (1973), as well as large articulated *Unio* sp. shells during the 1982 excavations. Above this horizon, shells were present scattered through the sediments and have been identified by Kerney (1971: see below). Immediately underlying the Lower Loam in the 1968–72 digs, shallow channel fills of sand, silt and fine gravel (75 cm maximum thickness) were found. These yielded both mammalian remains and Clactonian artefacts (Conway, 1971). Throughout the upper part of the Lower Gravels, clasts were coated with secondary calcite, the result of translocation of calcium carbonate from above.

This unit has been traced in a WNW–ESE direction through the Barnfield Pit and approximately W–E through Rickson's Pit (Dines, in Ovey, 1964). The overall width of the channel filled by the Lower Gravel is about 500 m, according to Waechter & Conway (1977).

Pebble counts were undertaken during the 1982 excavations (Appendix 2) and comprise an assemblage closely comparable to that for other Thames units identified in the vicinity. There is no doubt therefore that the deposits represent the river itself.

(b) Lower Loam

Resting on the Lower Gravel over much of the Barnfield Quarry is a unit that comprises predominantly brown sandy silt. It is generally 2–2.8 m thick in the quarry exposures and has been traced laterally for at least 200 m. However, trenches cut in the Alkerden Lane allotments showed the Loam was cut out to the west (Conway, 1973). It did not occur at Ingress Vale (Smith & Dewey, 1914) and was also absent to the east in Rickson's Pit (cf. Burchell, 1931, 1934b). The basal contact of the unit is undulating, apparently caused by the sediments blanketing bar and channel-like features with an amplitude of up to 1 m. This undulation appears to represent the topography of the underlying sediments. The Loam fines upwards from a basal light-brown sandy silt to a dark brown clayey silt in the upper 1 m. In addition, the basal silt includes horizontally discontinuous beds of fine to medium sand which vary considerably in thickness from narrow interlaminae in the basal 20–30 cm to individual bands up to 10–15 cm thick. These sand bands are often rich in shells. Conway (1973) reported one pebbly sand lens, 20 cm thick, the basal surface of which included 'a large number of roughly circular, often intersecting, depressions... ranging in diameter from 10–30 cm and ...up to 15 cm deep.' Although dismissed as unlikely by Conway (1973, p. 81), it is highly probable that these features are small potholes similar to those reported by Harding & Gibbard (1984) from Stoke Newington, and Bridgland *et al.* (1985) from the upper surface of the Lower Loam. The sand bands become thinner and less common upwards and are absent from the upper 1 m of the deposit.

The upper surface of the Lower Loam is at 25–27 m OD over much of the exposure. Irregularities in this surface, with an amplitude of up to 20 cm, which were observed after removal of the overlying gravel, were interpreted as scour features and include the small potholes referred to above. These features indicate that the surface was eroded over most of the area, but the discovery of footprints in some places demonstrates that locally erosion was extremely weak. The occurrence of a reddish zone in the uppermost part of the Loam, 50 cm to 1 m deep, has been confirmed as a temperate buried soil by Kemp (1985). This soil was formed when the surface of the Loam was exposed as a stable land surface and was subjected to carbonate leaching. Later gleying and mottling resulted from rising watertable levels and was followed by truncation and burial of the surface.

Clactonian artefacts are found throughout the Lower Loam, but are concentrated at certain horizons, interpreted to represent temporary desiccation events by Waechter & Conway (1977). Scattered vertebrate remains also occur (Sutcliffe, in Ovey, 1964).

(c) Middle Gravel, Upper Loam and 'Upper Gravel'

Although the Middle Gravel was originally recognised as a single unit by Smith & Dewey (1913, 1914), excavations following the discovery of the human skull fragments led Marston (1937) to subdivide the Middle Gravel sequence into

2.3 SWANSCOMBE MEMBER DEPOSITS

Lower and Upper subunits. The Lower Loam is overlain by 1.5–2 m of medium gravel in a sand-rich matrix, alternating with horizontally bedded sand in recent sections, although thicknesses of almost 5 m have been reported (Wymer, 1968). Thinning of the deposits may reflect proximity to the contemporary valley side. Although partially decalcified, shelly lenses occur locally at or to within 1 m above the base, according to Kerney (1971) and Conway (1973), both of whom give lists of Mollusca identified. The discovery of decalcified bone on the eroded Lower Loam surface supports the view that the lowest gravel, in part, represented a 'lag'. This subunit has yielded a very rich assemblage of Early Middle Acheulian pointed hand axes, all in slightly rolled condition (Wymer, 1968). Pebble counts from the Lower Middle Gravel from Barnfield and Galley Hill Pits given in Appendix 2 show no significant difference from those for the Lower Gravel.

The contact of the Upper Middle Gravel is undulating and erosional over much of the pit. In one place a substantial channel, cut through the Lower Middle Gravel, Lower Loam and much of the Lower Gravel, was reported by Marston (1937). The nature of this channel is discussed by Conway (1973), who concludes that it was probably a cavity formed by solution of the underlying Chalk bedrock, rather than a channel, the beds up to and including the Lower Middle Gravel having collapsed into the hollow. Infilling of the resulting cavity may have thus occurred during deposition of the Upper subunit. Which of these explanations is correct cannot be determined with certainty today: both explanations are plausible. Indeed, conceivably both are correct, in that during the collapse, fluvial erosion of pre-existing beds could have occurred. However, it is certain that the solutional collapse or channelling must have occurred during or immediately before deposition of the Upper Middle Gravel sands, because they apparently fill the cavity and are undisturbed (cf. Conway, 1973).

The deposits themselves are up to 4.5 m thick, and comprise a basal pebbly 'lag' including large flints, overlain by tabular cross-stratified, coarse to medium sand, which contains narrow pebbly sand bands up to 10 cm thick. This in turn gave way to a bed of ripple-bedded and small-scale trough cross-bedded sand with flint granules on the foresets. Palaeocurrent measurements from these sediments indicate a general ESE flow (mean direction 169°). In the uppermost 2 m the tabular cross-bedded sand units (each 10–15 cm thick) become interbedded with brown silty clay laminae ('drapes') up to 1 m thick, but thickening upward and eventually becoming dominant. At the same time, the sand bands decrease proportionally in thickness and eventually die out in the upper 85 cm of the section. Ground ice structures, including thermal contraction cracks, cryoturbations and faulting, have been seen in these sediments by Conway (1973), and Paterson (1940) recorded soliflucted gravel. These sediments represent a gradual, fining-upward transition into the 'Upper Loam' facies, culminating in a massive chocolate-brown silty clay, closely similar to reworked London Clay. This deposit contains angular flint pebbles irregularly distributed throughout. Waechter & Conway (1977) report drag folds beneath the Upper Loam, as well as inclusions of sand and pebble concentrations within the sediment, which they concluded was soliflucted.

Of the faunal remains, shells are rare, but vertebrate remains are quite common, although decalcified and very fragmentary (cf. Sutcliffe, 1964). The latter includes *Lemmus* sp., *Equus ferus* and *Canis* cf. *lupus* (Chapter 5). The human remains were recovered from about 60 cm above the base and were associated with numerous flakes and hand axes in sharp condition. Acheulian hand axes were also common in the basal 'lag' of the Upper subunit, whilst white patinated implements were commonly encountered immediately beneath the Upper Loam (Wymer, 1968). Waechter & Conway (1977) report the important find of musk ox *Ovibos* sp. from their 'Upper Gravel'.

The sequence presented above is essentially that at Barnfield Pit. However, as already emphasised, there are considerable lateral variations. According to Dewey (1932), at Rickson's Pit a basal coarse gravel, 1 m thick, which yielded an abundant Clactonian industry, was overlain by 3 m of sandy gravel with shells. This was succeeded by 60 cm of horizontally bedded sand and gravel and 1.3–2 m of current bedded sand and interbedded clay. The latter was separated from the bed beneath by a lens of 'rubble chalk' (Burchell, 1931, 1934b). In the upper sand and clay bed sharp Middle Acheulian hand axes have been found (Wymer, 1968). On the basis of the sedimentary sequence, the artefacts and the Mollusca (Kerney, 1971), there is little doubt that the basal coarse gravel is equivalent to the Lower Gravel at Barnfield Pit, whilst the overlying sediments are the extension of the Middle Gravels.

Towards the west, the exposure at Ingress Vale again differed from that at Barnfield Pit (Stopes, 1903; Smith & Dewey, 1914). At the former the deposits rested on Chalk at 24–25 m OD (Fig. 9). Red, occasionally contorted sand and gravel, 30–60 cm thick, generally occurred at the base and were overlain by 60 cm to 2 m of sand containing abundant Mollusca. Resting on the shelly sand with a marked erosional base was current bedded sand, which locally reaches 3 m in thickness, above which a bed of gravel *c.*50 cm thick was present. The sections are capped by 1–1.5 m of so-called 'stony loam', the descriptions of which suggest that it is probably the solifluction deposit of reworked London Clay seen in the Barnfield sections.

There is some disagreement between authors over the relationship of the Ingress Vale sediments to those at Barnfield Pit. Smith & Dewey (1914) concluded that the shelly sand and underlying gravel were equivalent to the Barnfield Lower Gravel and Lower Loam (cf. Wymer, 1968), whereas Kerney (1971) concluded that the Mollusca represent a latter period, since the fauna is similar to that from the Lower Middle Gravel. He also mentioned finds of Acheulian material from the shelly sands in unpublished excavations by himself and Sieveking. Although this dilemma could not be resolved from field study during this project, the answer may be that both authorities are correct. The basal gravel might be a partial equivalent of the Barnfield Lower Gravel, whilst the shelly sands are presumably later, as Kerney (1971) shows. These sands could represent the flow channel of the river spanning the period when the Lower Loam surface was exposed to subaerial weathering, and the period immediately afterwards when the Lower Middle Gravel began accumulating. From this conclusion, it follows that the current bedded sands that overlie the shelly beds at Ingress Vale are equivalent to the widespread Upper Middle Gravel unit (Fig. 9).

Finally, north of the Barnfield Pit, the bedrock surface rises to 27.5 m OD where it is exposed in the chalk workings on the east side of Crayland's Lane (TQ 604746; Fig. 1). Here the deposits comprise 1.5 m of medium gravel in a sand matrix, the basal gravel being very coarse. This is overlain by 2.3 m of horizontally bedded medium gravel and sand. Cryoturbated orange brown clay-rich sandy gravel, 50 cm to 1.2 m thick caps the sections. This description is closely comparable to that given by Smith & Dewey (1914) when the pit was opened. Today the sequence is still visible in the overgrown faces. Palaeocurrent measurements indicate a flow towards the north-east, and a pebble count from the gravels is given in Appendix 2. Evidence of solutional collapse of the underlying bedrock, causing local thickening of the gravels, can be seen in several places. The Chalk rises to the ground surface at the eastern end of the quarry, cutting out the Pleistocene deposits. The boundary of the deposits trends ENE across the workings and therefore diverges from the eastward line of the Barnfield deposits such that the two sediment suites are separated by the bedrock high. This evidence, together with the markedly different sediment facies preserved at this site, suggests that the deposits are not equivalent to any of those at the neighbouring Barnfield Pit (*contra* Smith & Dewey, 1914, and subsequent workers, cf. Wymer, 1968), but are the correlative of the younger, Orsett Heath Gravel (see below), the boundary between these two groups of sediments being as suggested by Oakley (in Ovey, 1964; Fig. 2).

In conclusion, the following sequence of events seems to be represented at Swanscombe:

1. Downcutting to the surface underlying the Lower Gravel.
2. Accumulation of the Lower Gravel, followed by the Lower Loam.
3. Exposure of Lower Loam surface to subaerial weathering, giving rise to soil development. Contemporary channel flow and deposition at Ingress Vale.
4. Rejuvenation of the river causing deposition of gravel and sand (Lower Middle Gravel), followed by solutional collapse in Barnfield Pit.
5. Rejuvenation continues with widespread deposition of sands and gravel (Upper Middle Gravel).
6. Infill of channel recorded by gradation from sands into sand/silt interbeds and ultimately clays (Upper Loam), under periglacial conditions.
7. Solifluction of clayey gravel, probably derived from London Clay above site to south.

2.4 Orsett Heath Gravel

Because they comprise the highest coherent terrace aggradation in the valley, the deposits of this unit were originally referred to as the 'High or 100 foot Terrace' (Hinton & Kennard, 1901, 1905, 1907; Pocock, 1903; Smith & Dewey, 1914). In his memoir on the London District, Woodward (1909) proposed a modified scheme in which he suggested the term '2nd terrace' for the unit. However, mapping of the Middle Thames region led the Geological Survey to define terraces from type localities; the terms Boyn Hill Terrace and Boyn Hill Gravel (based on a stratotype in Maidenhead) were used to identify the then highest terrace (Bromehead, 1912; Dewey & Bromehead, 1915). In subsequent mapping downstream, the Boyn Hill Terrace has been identified throughout the Lower Thames Valley. The implement and fossil-rich Swanscombe deposits (see above) were also included in the Boyn Hill on altitudinal grounds by Dewey &

Bromehead (1915), and this correlation has remained virtually unquestioned until recently. Indeed the Swanscombe sequence was considered as the Lower Thames' stratotype of the '100 foot Terrace' (Waechter & Conway, 1977) until it was realised, from work in the Middle Thames and that presented here, that the Boyn Hill and the Lower Thames equivalent deposits differ considerably in character from those at Swanscombe (Gibbard, 1985; Gibbard, Whiteman & Bridgland, 1988; Bridgland, 1988a). Gibbard (1985) identified the Boyn Hill Member as a cold climate aggradation, younger than that at Swanscombe, but abutting the sediments at this site (cf. Oakley, in Ovey, 1964; see above).

Dewey (1932) and later Wooldridge & Linton (1955), discussing the recognition by Holmes (1892, 1894) that the gravels of this unit overlie the chalky till at Hornchurch, suggested the aggradation must post-date the Anglian glaciation. This evidence for a long time provided a vital link with the glacial sequence of eastern England, and, since the gravels post-date the Hoxnian sediments at Swanscombe, reinforces the view that the member is of Wolstonian age (Gibbard, 1985).

Because of the confusion and mis-use of the term Boyn Hill, a new member name has been proposed for this unit in the Lower Thames Valley. This term, 'Orsett Heath Gravel', based on a stratotype at South Orsett (TQ 628810: 26 m), was originally proposed by Bridgland (1983a, 1988a) and replaces a preliminary term 'Fairlop Gravel' (Gibbard, 1985) which has been shown to be based upon younger, Lea/Thames confluence deposits of the Corbets Tey Member (Gibbard, Whiteman & Bridgland, 1988; see below). The stratotype defined here is different from that originally proposed by Bridgland (1983a, 1988a) at TQ 668803. This is because the deposits in Bridgland's section are here correlated with the Dartford Heath Gravel (see above).

The type section comprises a large disused gravel pit that exposes 4.0 m of gravel and sand, the upper 1.5 m of which is much disturbed by cryoturbation. The deposits consist of horizontally stratified 10–15 cm thick units of fine to medium gravel in a sand matrix, interbedded with tabular cross-bedded sand in bands and shallow channels 5–15 cm thick. Palaeocurrent measurements from the sand indicate flow towards the north-east. The pebble count from a gravel sample is given in Appendix 2. In places, the gravels and sands are overlain by 50 cm of brown silty sand, which includes some scattered frost-shattered flint pebbles in the basal part (section 2.10).

The Orsett pit occurs adjacent to the A13 road, a section along which is shown in Fig. 7. In this section the gravels rest on Tertiary bedrock. The relationship of this member to the higher Dartford Heath Gravel is visible in this section. Here the base of the former is $c.8.25$ m lower than the latter.

The Orsett Heath area is underlain by an extensive spread of these deposits, as noted by Dewey *et al.* (1924). The deposits are also exposed at Chadwell St. Mary (TQ 651783: 27 m) where they are generally 5 m thick, but rest on an undulating base of Thanet Sand (see Fig. 47 below). Locally, channels up to 2 m deep are cut into the bedrock. The deposits comprise massive gravel units 20–50 cm thick, separated by occasional lenticular interbeds of tabular cross-bedded medium to coarse sand, 5–30 cm thick. An eastward palaeoflow direction is indicated by the gravel imbrication and palaeocurrent measurements, and a pebble lithological count is given in Appendix 2. This and the old section at Sockett's Heath (*c.*TQ 629794: 25 m) were also seen by Dewey *et al.* (1924)

2.4 ORSETT HEATH GRAVEL

during the geological mapping. At the latter, 4.5 m of sand and gravel were present, the lower half consisting of current bedded sand. A number of erratics and Palaeolithic artefacts were collected from this site at the turn of the century by Hinton & Kennard (1901). The extensive spread of gravels of this unit are cut out at West Tilbury by the lower, Mucking Gravel, but a small outlier may be present near Gun Hill (TQ 656780: 25 m) where the sandy gravels are 4 m thick and rest on Thanet Sand.

Immediately upstream of the Orsett Heath spread is that north of Grays and South Stifford. This patch has been highly dissected by chalk quarrying. There are, however, some large but generally inaccessible exposures in the quarry faces. For example, at Grays Chalk Quarry (TQ 605793: 33 m) 1.75 m of stratified gravel and sand overlie Thanet Sand. Penetrating the latter is a vertical pipe-like structure, 4 m deep and 20 cm wide, containing partially stratified gravel and sand, and a Thanet Sand block. Similar extensive sections were visible adjacent to Pilgrim's Lane, South Stifford (TQ 594787–596787: 33 m), where 2.5–3 m of stratified medium gravel and sand overlies Thanet Sand. The gravels are predominantly horizontally bedded and are separated by tabular cross-bedded sand units up to 50 cm thick. Cross-bedding measurements from the sands give a palaeocurrent towards the north-east. Again the gravel base is generally almost horizontal, but locally small potholes occur, and in the southern part of the exposure, a large solution hollow in the Chalk bedrock that had effected both the Thanet Sand and the Pleistocene deposits was infilled by 2 m of gravels and sands. This unit is again seen in the neighbouring quarry at TQ 592787 (34 m). Pebble counts from samples collected from these sites are listed in Appendix 2.

The sediments on the south side of the river are exposed at Crayland's Lane, Swanscombe (TQ 604746; Fig. 9), already described above (section 2.3) as abutting the upper part of the Swanscombe sequence. The deposits of this unit can be traced westwards to Horns Cross (TQ 581745: 31 m) where 3 m of horizontally stratified medium gravel is exposed. Both imbrication of the clasts and palaeocurrent measurements indicate flow towards ENE. At Stone Cross (TQ 567742: 31 m), 1.5 km further west, very similar deposits were exposed in 1983. Here the medium gravel with a granule-dominated matrix occurs in horizontal units 35–60 cm thick, interbedded with narrow lenses of tabular cross bedded sand. A total thickness of 5.8 m of deposits was exposed, here resting on bedrock chalk and they abut the Dartford Heath Gravel immediately to the west and south.

This unit is next encountered as two adjacent, isolated patches at Crayford. The first, around St Paulinus Church (TQ 512751: 33 m) is at least 1.5 m thick in the churchyard. At the second, at Manor House Barnehurst (TQ 512755: 31 m), a temporary exposure revealed 2 m of stratified pebbly sand, the imbrication in which indicated a current flow towards the south-east. An exposure in this spread showing gravel resting on Thanet Sand at 27.5 m OD was reported by Chandler (1914). Pebble counts from both localities, given in Appendix 2, are rich in Greensand chert. This was also noticed by Dewey et al. (1924, pp. 5–6). The same authors also mapped a small patch of gravel and sand at Mascal Hill (TQ 494727: 33 m) in the Cray valley that they considered as a possible tributary equivalent of this member. On the basis of its altitude, this patch could represent such an equivalent, and, if so, implies the contemporary existence of the Cray.

No further spread of the Orsett Heath Member is seen upstream on the southern side of the river in the area. However, on the north side it is well represented, and is next encountered at Moor Hall, Aveley (TQ 557812: 33 m), where sections in the overgrown pit show 4.5 m of gravel and sand resting on London Clay. As elsewhere, the gravels occur as subhorizontal bands 60–70 cm thick interbedded with 10–15 cm thick sand beds. Palaeocurrent measurements unquestionably indicate deposition by flow towards the southwest, as previously noted by earlier writers (e.g. Wiseman, 1978; Bridgland, 1988a). This conclusion is interpreted as resulting from the river flowing westwards at this point, adjacent to the high chalk ridge of the Purfleet anticline (see below).

In the Havering area, fragments of 'Boyn Hill' gravel are mapped on the interfluves between small southward flowing tributaries, the Rom and Ingrebourne. In Upminster and Hornchurch these spreads occur between 34 m and 38 m, the major section being in Hornchurch railway cutting (TQ 54708737: 36 m). As already mentioned, recent re-excavation of this classic section by English Nature (Bridgland, Cheshire, Gibbard, Harding & Whiteman, unpublished) confirmed Holmes' (1892, 1894) observations (Fig. 5). Here the gravels and sands are 3.8 m thick and again comprise horizontally stratified medium to fine gravel beds with narrow stratified sand bands between. A post-depositional ice wedge cast penetrated and disturbed the gravels to a depth of 3 m from the surface. Pebble count results are presented in Appendix 2. Similar sections were recorded in excavations at the Dell, Hornchurch (TQ 546868: 34 m), but here the till was cut out, the gravel and sand resting directly on London Clay. The large patches of gravel and sand mapped as 'Boyn Hill' beneath Heath Park and Gidea Park, north-west of Hornchurch, were not exposed during this investigation, but they occur in the height range of the Orsett Heath Member. The occurrence of 2.4 m of gravel and sand at the appropriate altitude is confirmed by boreholes for a railway bridge at Gidea Park (TQ 531895: 34 m). The gravel and sand mapped at Ardleigh Green (TQ 538896: 35 m) is also confirmed by boreholes and is 3.75 m thick.

According to the geological maps (sheets 256, 257) extensive 'Boyn Hill' gravel spreads underlie the Fairlop–Gants Hill area on the Rom–Roding interfluve, and the Wanstead area on the adjacent Lea–Roding interfluve. Both these spreads are certainly underlain by considerable thicknesses of gravels and sands, but they occur at altitudes of 25–34 m OD. By comparison with the heights of the Orsett Heath, these spreads appear to be too low (see Fig. 21 below). They are therefore assigned here to the Corbets Tey Member (see below).

No further outliers have been found upstream in the area at the appropriate elevation for inclusion in this unit. However, at Islington (TQ 313837:42 m), and south of the river at Herne Hill (TQ 324747: 43 m), deposits occur that were assigned by Gibbard (1985) to the Boyn Hill Gravel of the Middle Thames. The former was proved in boreholes (Fig. 10) where they are 7.5 m thick and lie on London Clay at 31.5 m OD. Although lack of exposure prevented correlation using lithology, altitudinally these deposits may equally be included in the Orsett Heath Gravel. If correct, it therefore appears that the Orsett Heath Gravel is indeed the downstream equivalent of the Boyn Hill Gravel (cf. Gibbard, 1985; Gibbard, Whiteman & Bridgland, 1988; Bridgland 1988a).

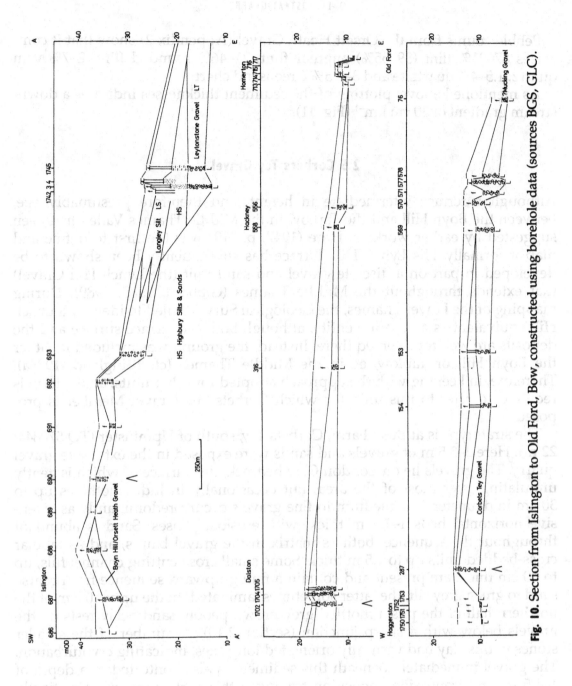

Fig. 10. Section from Islington to Old Ford, Bow, constructed using borehole data (sources BGS, GLC).

Pebble counts from the Orsett Heath Gravel (Appendix 2) show that it comprises 87–91% flint (39–55%) angular flint, 21–46% rounded flint, 2–7% vein quartz, 1.5–4% quartzite and 3–6.5% Greensand chert.

As mentioned above, plotting of the sediment thicknesses indicates a downstream gradient of 30 cm km^{-1} (Fig. 11).

2.5 Corbets Tey Gravel

Although a terrace intermediate in height, and therefore presumably age, between the Boyn Hill and the Taplow in the Middle Thames Valley had been suggested by earlier workers, Hare (1947, p. 319) was the first to define and map it formally. His Lynch Hill Terrace has subsequently been shown to be developed in part on a discrete gravel and sand unit (the Lynch Hill Gravel) that extends throughout the Middle Thames (Gibbard, 1985, 1989). During mapping of the Lower Thames, the Geological Survey failed to identify a Lynch Hill equivalent as a separate entity, although both the terrace surface and the deposits are well represented there. Instead, the ground was included in either the Boyn Hill or Taplow, as in the Middle Thames (cf. Bridgland, 1988a). Therefore, in keeping with the approach adopted for other units, a new term is required to refer to this unit, for which Corbets Tey Gravel Member is proposed.

The stratotype is at Bush Farm, Corbets Tey, south of Upminster (TQ 570844: 22 m). Here 3–3.5 m of gravels and sands were exposed in the extensive gravel quarry. The gravels lie on London Clay bedrock, the surface of which is gently undulating over most of the area, but occasionally includes potholes up to 30 cm in diameter. The medium to fine gravels occur predominantly as extensive horizontal beds 1–1.5 m thick, with erosional bases. Sand is abundant throughout the sequence, both as matrix in the gravel bands, and as tabular cross-bedded units up to 1.5 m thick. Some small cross-cutting channel fills, up to 20 cm thick, are present and contain a fining-upward sequence from coarse sand to green grey silt, the latter sometimes laminated. In the upper 0.5 m of the northern half of the pits a mottled grey-brown pebbly sandy clay rests on the gravels below with a sharp junction (section 2.10). A number of the smaller stones in this clay had vertically orientated long axes, indicating cryoturbation. The gravel immediately beneath this sediment is also contorted to a depth of 1–1.5 m, and iron oxide deposition has given this part of deposit a distinctly orange-brown colour. This possibly represents a palaeosol.

A number of exposures occur in the vicinity and show a remarkably similar sequence: in particular, those at Gerpins Lane (TQ 550833: 15.5 m) and Buckles Farm, South Ockendon (TQ 594813: 21 m). At South Ockendon, only 70 cm of stratified gravel, overlain by c. 1 m of brown silty clay with occasional flint pebbles, were visible, because the workings were flooded. At the former, the gravels and sands were 3 m thick. Whitaker (1889) reported gravel and sand beneath 'brickearth' c.1 km south-west of Cranham Church (TQ 567852: 22 m).

The deposits are next seen in North Ockendon at two large quarries: Baldwin's Pit (Cawood) (TQ 571826: 19.5 m), and Baldwin's Pit (Redland), near Dennises Cottages (TQ 578838: 22 m). At the latter the sequence was 4.5 m thick overall, the deposits resting on London Clay. The deposits comprise horizon-

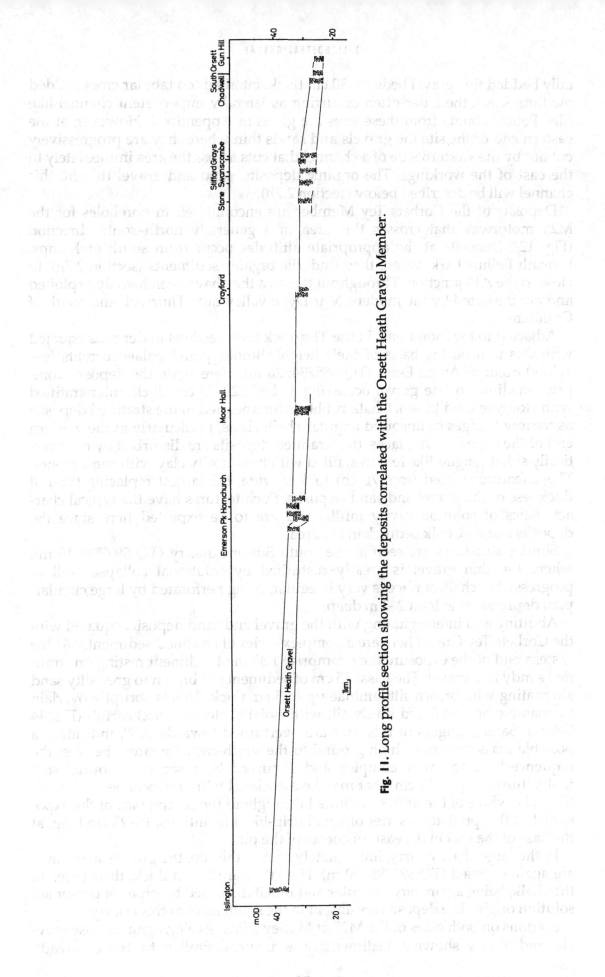

Fig. 11. Long profile section showing the deposits correlated with the Orsett Heath Gravel Member.

35

tally bedded fine gravel beds 20–30 cm thick, interbedded tabular cross bedded medium sand, the latter often occurring as laterally impersistent channel-like fills. Pebble counts from these sites are given in Appendix 2. However, at the eastern end of the site the gravels and sands thin where they are progressively cut out by the western side of a channel that cuts across the area immediately to the east of the workings. The organic deposits, sand and gravel that fill this channel will be described below (section 2.7h).

Deposits of the Corbets Tey Member are encountered in boreholes for the M25 motorway that crosses this area in a generally north–south direction (Fig. 12). Deposits at the appropriate altitudes occur from south of Kemps, beneath Belhus Park, where they underlie organic sediments (section 2.7g), to close to the A13 junction. Throughout the area they have been heavily exploited and are dissected by the modern Mar Dyke valley near Thurrock and south of Cranham.

Adjacent to the motorway, in the Thurrock area, sections in deposits equated with this unit on the basis of their height, lithology and palaeocurrents (see below) occur at Arena Essex (TQ 585794: 20 m). Here again the deposits comprise medium to fine gravel occurring as beds 20–25 cm thick, interstratified with elongate sand lenses. Chalk rubble is interbedded in the stratified deposits as narrow wedges of unsorted angular chalk clasts, particularly at the western end of the exposure. In places, the stratified deposits are disturbed by near vertically sided tongue-like features, filled with brown silty clay with some stones. These features varied from 70 cm to 4 m wide, the largest replacing the full thickness of the gravel and sand deposits. Such features have the typical characteristics of solution cavity infills, and are to be expected here since the deposits rest on Chalk bedrock in the area.

Similar structures are seen at the South Stifford quarry (TQ 593794: 19 m), where the thin gravel is greatly disturbed by solutional collapse (still in progress); the chalk surface is very irregular, being perforated by large circular-plan depressions at least 2–3 m deep.

Abutting and interdigitating with the gravel and sand deposits equated with the Corbets Tey Gravel here are a complex series of stratified sediments. At the eastern end of the exposure they comprise 1.35 m of sediment resting on stratified sandy fine gravel. The basal 35 cm of sediment is a brown to grey silty sand alternating with brown silt laminae up to 5 cm thick. This is abruptly overlain by massive brownish-red sandy silt with isolated stones ('brickearth') (Fig. 44 below). Sand stringers in this unit are overturned towards 295°, indicating a possible mass flow from high ground to the south-east. Towards the west the sequence becomes more complex and disturbed by a series of normal step faults (throws up to 55 cm) that may be associated with the local bedrock solution. The whole of the unit is involuted throughout the central part of the exposure, but this predates a series of small drip-like involutions, 15–25 cm long, at the base of the unit in the eastern corner of the pit.

In the large chalk quarry, immediately east of the site, the gravels and sands are again exposed (TQ 596795: 20 m). Here they are 2–3 m thick, their base, on the chalk, being again very irregular and much disturbed by pipes of presumed solution origin. The deposits are absent in the south faces of this quarry.

Sections on both sides of the M25 at Marley's Pits, Aveley, again expose gravels and sands showing sedimentary structures similar to those already

2.5 CORBETS TEY GRAVEL

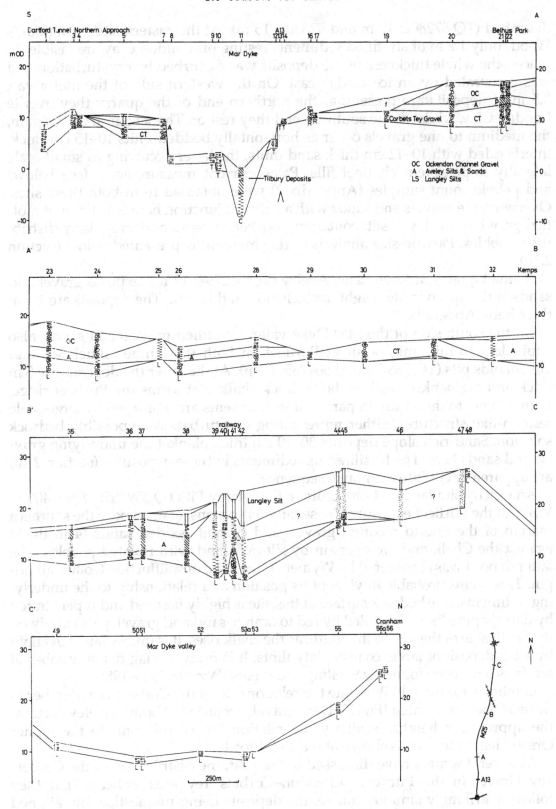

Fig. 12. Section along the M25 motorway from the northern approach to the Dartford Tunnel, Purfleet to Cranham, constructed using borehole data (source: ERCU).

described (TQ 572802: 17 m and 578802: 15 m). At the eastern site (Hangman's Wood) only 1.2 m of stratified sediments resting on London Clay are visible. In places, the whole thickness of the deposits was disturbed by cryoturbation and was penetrated by an ice wedge cast. On the western side of the motorway, 3.5 m of deposits are present; at the northern end of the quarry they overlie London Clay, whilst at the southern end they rest on Thanet Sand. Once again, the medium to fine gravels occur as horizontally bedded units 10–15 cm thick, interbedded with 10–12 cm thick sand units, the latter occurring as small-scale laterally impersistent channel fills. Palaeocurrent measurements (see below) and pebble count samples (Appendix 2) were obtained from both these sites. Overlying the gravels and sands with an abrupt junction here is 1–1.5 m of mottled grey-brown clayey silt, containing pockets of sand and irregularly distributed pebbles. Particle-size analysis of this material is presented below (section 2.10).

A small quarry at Love Lane, Aveley (TQ 562794: 17 m), exposes gravel and sands at the appropriate height for inclusion in this unit. The deposits are 1.5 m thick here (Appendix 2).

On the south side of the Mar Dyke valley, stratified gravels and sands, also equivalent to this unit, occur at Purfleet, in both the famous Bluelands and Greenlands pits (TQ 568787 and 564785: 17 m). At the latter the deposits are 5 m thick and are banked against the bedrock chalk that forms the Purfleet ridge, immediately to the south. In parts, these sediments are disturbed by large-scale festoon-like structures, either representing cryoturbation or possibly bedrock solution. Sand-rich slope deposits 30–40 cm thick blanket the underlying gravels and sands here. The fossiliferous sediments in these exposures (section 2.10) are apparently cut into the gravel sequence.

An exactly similar situation occurs in the Botany Pit (TQ 559785: 17 m), 400 m WSW of the Purfleet exposure. These famous sections again expose the southern margin of the fine to medium gravels and current bedded sands (4 m thick) against the Chalk and are overlain by silt-rich sand with isolated pebbles. The latter deposit was considered by Wymer (1968) to be of solifluction (colluvial) origin. This seems probable in view of its position and relationship to the underlying sediments. The bedrock surface at this site is highly fissured and is penetrated by narrow pipe-like tubes filled by red to orange sand and gravel, particularly on the north side of the pit. To the south, as the chalk rises, it is immediately overlain by a lag deposit of large, coarse, platy flints. It is from this lag that a number of artefacts have been found by Snelling and others (Wymer, 1968, 1985).

Boreholes for the A13 illustrate the relationship of the Corbets Tey Member to those in the Aveley area (Fig. 7). Here gravels recorded at South Aveley occur at the appropriate height. Similarly the relationship of this unit to the higher Orsett Heath Member is shown at North Stifford.

As several writers have discussed in the past, the distribution of the Corbets Tey Gravel in the Purfleet–Ockendon–Corbets Tey area indicate that they follow a strikingly sinuous course, the deposits being unquestionably aligned initially eastwards as far as Corbets Tey, but turning sharply southwards to Ockendon, and continuing towards the south-west to Purfleet where they are banked against the Purfleet anticlinal chalk ridge. This course, recognised originally by Wooldridge & Linton (1955), has been discussed by Bridgland (1988a) and Wiseman (1978), the latter with reference to the Orsett Heath Gravel at

Aveley (see above). On the basis of the palaeocurrent measurements obtained from the exposures during this study, there can be no doubt that this 'double-bend' course was occupied by the Thames during deposition of the Corbets Tey Gravel (*contra* Wymer, 1968; Palmer, 1975).

Downstream of the Purfleet gap this unit is less well represented. It is next encountered at the Globe Pit, Grays or Little Thurrock (TQ 624783: 15 m) where it reached a maximum of 2 m thick and rested on Thanet Sand. Sections at this site have been described by Wymer (1957, 1968), who records numerous Clactonian flakes from this deposit.

Projection of the gradient of 35–40 cm km^{-1} (see Fig. 21 below) indicates that the small outlier at Barvill's Farm (TQ 682778: 14 m) should be included with this unit. In addition, Bridgland (1988a) suggests that this member is also represented in a gravel spread beneath Stanford-le-Hope (TQ 694825: 17 m) and Corringham (TQ 710825: 17 m). At neither of the latter sites were exposures available for study; it was not therefore possible to confirm the actual thickness of the member or whether later slope deposits overlay the gravels here.

Upstream of the stratotype, the gravels continue as a dissected spread south of Hornchurch, where they abut the Orsett Heath Member near St George's Hospital, to as far as Romford. The deposits were visible in a temporary ditch at Crown Farm (TQ 495885: 18 m), where they comprised uniform horizontally stratified gravel in a sand matrix. The deposits are over 1.75 m thick here and are overlain by 15 cm of grey pebbly clay.

To the west, much of the Redbridge area is underlain by deposits that can be assigned to this member, including that beneath Goodmayes, north Ilford, Gantshill, Fairlop, Little Heath and Chadwell Heath. The latter areas are included in the Boyn Hill Terrace spread according to the Geological Survey map (sheet 257). However, as noted above, the ground cannot be included with the Boyn Hill equivalent, the Orsett Heath Gravel, because it is uniformly too low. At first sight it appears to be too high for inclusion in the Corbets Tey Member, but field inspection confirms mapped evidence that the spreads are continuous; the deposits here and on the Roding–Lea interfluve, appear to represent a fill formed at the Roding–Lea–Thames confluence.

The substantial gravel workings at Fairlop (TQ451895: 28 m) exposed the deposits. The sediments comprise 6 m of stratified gravel and sand resting on an undulating surface of London Clay. At the base, medium gravel, in places cross-stratified, 1.5 m thick is overlain by tabular cross-bedded sand with a few pebbles on the cross laminae. In turn, this is overlain by interbedded fine gravel with a coarse sand matrix, and cross-stratified gravel, 2 m thick. The upper part of the unit is a very sandy, horizontally bedded gravel. The upper 1.5 m is strongly reddened and iron stained, the matrix becoming very clay rich. The gravels were sampled at both 2 m and 4 m above the base (Appendix 2), and palaeocurrent measurements indicate a flow towards ESE. The deposits are capped by 1 m of chocolate-brown mottled grey silty clay with isolated pebbles that rests with marked unconformity on the deposits beneath. Laterally equivalent deposits were encountered in boreholes for the proposed M12 motorway (Fig. 13). In this section the gravels reach a maximum of 7 m and are overlain by pebbly clay up to 7.5 m thick. The latter is presumably the same as that seen in Fairlop Quarry, and is probably a mass flow deposit derived by solifluction from the neighbouring London Clay slopes adjacent to this spread. A similar

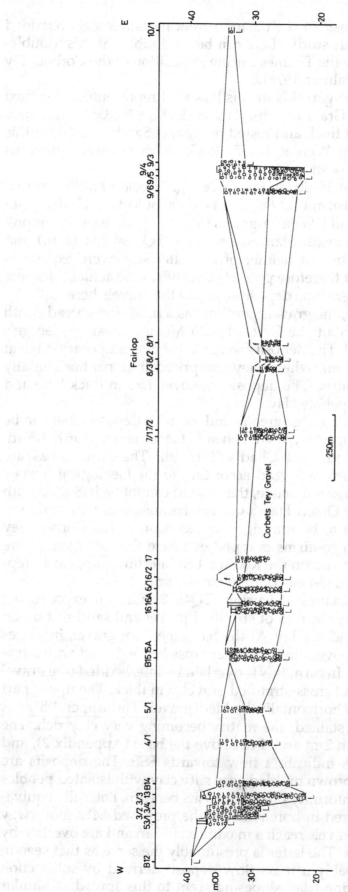

Fig. 13. Section along the proposed route of the M12 motorway from Woodford to Fairlop, constructed using borehole data (source: ERCU).

sequence occurs at Barkingside (TQ 446896: 28 m), where boreholes confirm that up to 1.5 m of brown clayey sand and gravel overlie 2.2 m of gravel and sand that rests, in turn, on London Clay.

Remnants of a possible Roding Valley equivalent of these deposits (mapped as Boyn Hill) were recognised at Chigwell, Loughton and Stapleford Tawney by Dines & Edmunds (1925).

According to Wymer (1985), Palaeolithic artefacts have been found at a number of sites in the Fairlop–Rush Green spread. These include Barkingside (TQ 445896), Clayhall, St. Swithin's (TQ 420895), Gants Hill (TQ 425876) and Redbridge (TQ 422882 and 419888). The remains of *Bos/Bison* have been reported from the St Swithin's site gravels by Hinton (1900).

The same author described a site near Wanstead Park, on the Lea–Roding interfluve, where 5 m of gravels and sands were exposed (*c*.TQ 412878: 25 m). A 'portion of a skeleton of *Equus*' was found here at a depth of *c*.3.5 m. The spread of deposits on this interfluve are a continuation of those already described. The deposits are clearly seen in the borehole sections for the Hackney–M11 link road (Fig. 14) and the Finsbury Park–Walthamstow–Roding Valley route (Fig. 15). The former indicates a spread of gravel and sand, up to 6.5 m thick, between Leytonstone Hospital and Wanstead Common, and confirms that the deposits thin north-westwards along Whipps Cross Road. The latter encounters the gravel and sand at the appropriate altitude at Snaresbrook. Sections in this area are rare, but 3 m of red-stained fine to medium gravel and sand, with interbedded tabular cross-bedded sand lenses can be seen in the banks of the large pond in Snaresbrook Park (TQ 393889: 31 m). Palaeocurrent measurement indicated a flow towards ESE, and a pebble count sample from this gravel is given in Appendix 2. The gravel and sand was also sampled from a borehole opposite Wanstead Station (TQ 406882: 29 m).

West of the Lea Valley, altitudinally equivalent gravel and sands occur in the north-east to east London area. This spread, mapped as 'Taplow' on sheet 256, in fact comprises a complex sequence of sediments, particularly in the Stoke Newington–Hackney Downs district. The geological sequence in this area has been the subject of periodic investigation over the last century in connection with the Palaeolithic artefacts and associated fauna discovered during house building. W. G. Smith (1879, 1894) reported the discoveries, which were particularly associated with sediments immediately north of Stoke Newington Common. Other writers such as Greenhill (1884), Whitaker (1889), Woodward (1909), Warren (1912, 1942) and Wymer (1968) have considered the area and generally restated Smith's ideas concerning the possible existence of an occupation surface or 'floor' in 'brickearth' type sediments. Recent reinvestigation (by the author) has greatly clarified the geological sequence in the district and thereby the context of the archaeology and palaeontology (section 2.7j).

On the basis of the borehole data collected, Corbets Tey Gravel occurs over a triangular-shaped area from Homerton to the Fleet Valley (Holborn Viaduct) and Clerkenwell to Dalston. Although much of this area has been greatly disturbed for centuries by the extraction of gravel and 'brickearth', it is still possible to determine the occurrence of the gravel and sand spreads with some accuracy. The sections (Figs. 10, 16, 17, 18, 19, 20, 34), constructed in an east–west and north–south grid, illustrate the relationships of the sediment units in the Stamford Hill–South Hackney area.

2 LITHOSTRATIGRAPHY

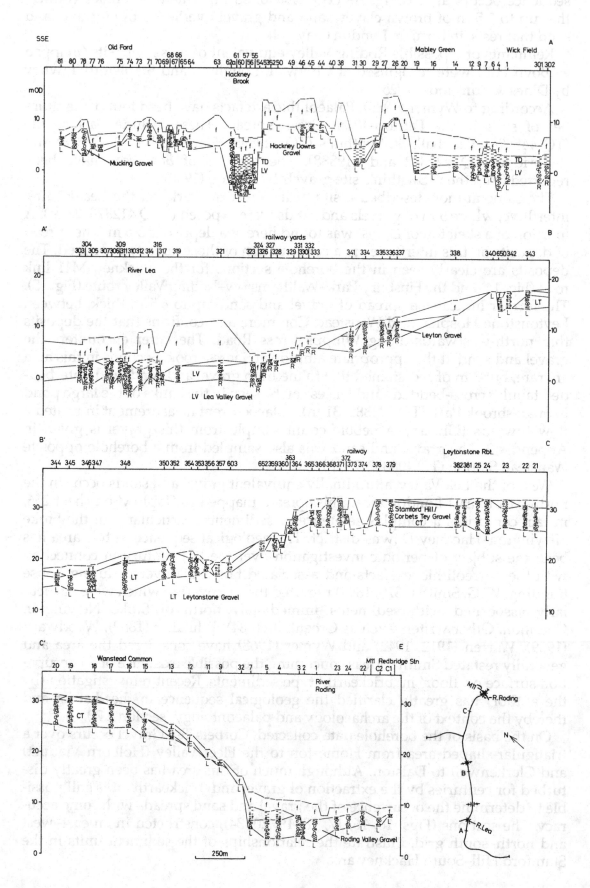

2.5 CORBETS TEY GRAVEL

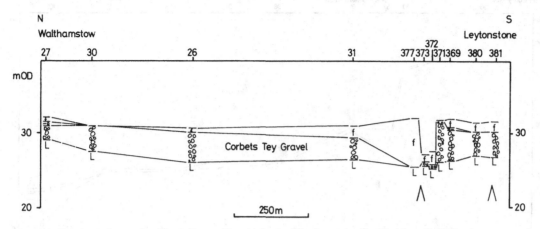

Fig. 16. Section along the East Cross Route from Hackney to Redbridge (M11), constructed using borehole data (source: GLC).

Beneath Hackney Downs to Islington, organic sediments separate the gravels into two subunits (sections 2.6, 2.7j). Abutting these spreads, immediately to the south, are gravels beneath Hackney Downs Station and Dalston. Sections in this area, particularly in the cuttings for the North London Line, were also observed by Whitaker (1889), whose descriptions confirm the borehole records. On the basis of these records it is not clear whether the gravels underlying the organic sediments are the lateral equivalent of those further south or not. From their altitude and distribution, this seems unlikely although not impossible; a bedrock step near Hackney Downs Station possibly indicates the break to a lower (main) aggradation.

Whether or not this is so, the wide spread to the south already mentioned can be seen in Figs. 16, 17, 19, 20, 34. Although dissected by younger stream valleys, the gravels and sands reach 9.2 m thick in the Clerkenwell–Shoreditch area, and occur between 20 m and 10.75 m OD. Numerous palaeoliths have been found from these deposits, according to Wymer (1968). The relationship of this member to the lower Mucking/Taplow Gravel can also be seen near Shoreditch and South Hackney, near Victoria Park. Indeed this line is very close to that mapped by the Geological Survey (sheet 256) as the 'Taplow–Floodplain Terrace' junction. The separation of these two members in the City of London and immediately to the north is difficult without borehole data, because of the close coincidental heights of these units. This results from the development of the Lea–Thames confluence in the area (see below).

West of the Fleet Valley, the spread continues at precisely the same height beneath Holborn. Bromehead (1925) and Dewey (1926) considered these deposits to be part of the same terrace unit that was recently assigned by Gibbard (1985) to the Middle Thames' Lynch Hill Gravel. Downvalley gradient supports this conclusion and demonstates that the Corbets Tey and Lynch Hill Gravels are lateral equivalents.

The Corbets Tey Gravel can therefore be followed from central London downstream to near Tilbury (and beyond: Bridgland, 1988a). Plotting of the sediment thicknesses indicates a downstream gradient of 35–40 cm km^{-1} (Fig. 21). The gravel comprises 84–94% flint (39–74% angular flint, 15–45% rounded flint), 3–10% vein quartz, 0–3% quartzite and 0–4% Greensand chert (Appendix 2).

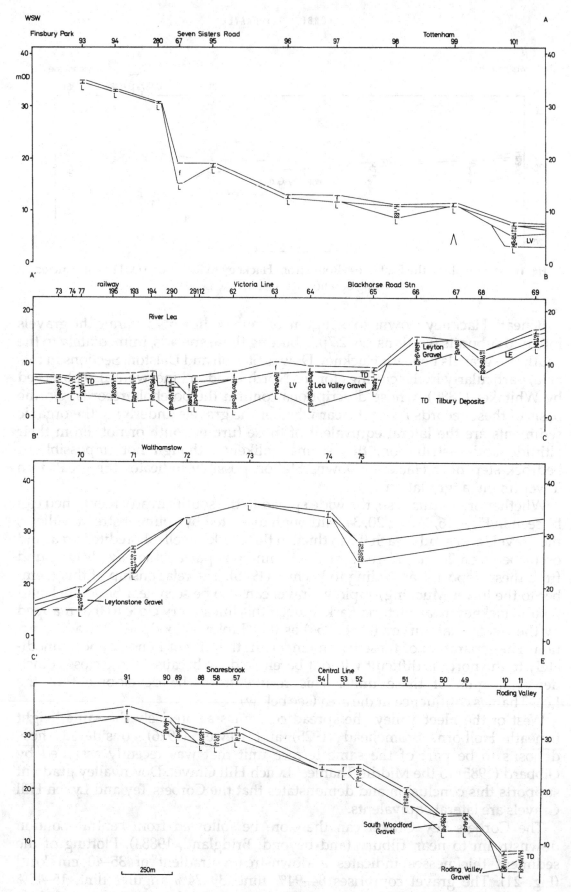

Fig. 15. Section for the proposed North Circular Road route from Finsbury to the Roding Valley junction with the M11, constructed using borehole data (source: GLC).

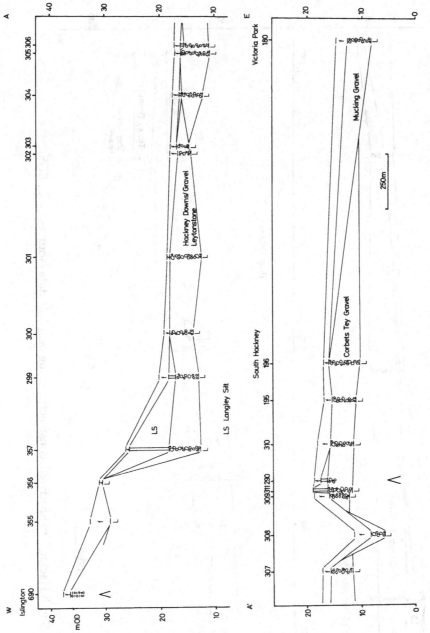

Fig. 16. Section from Islington to South Hackney, constructed using borehole data (sources: GLC, BGS).

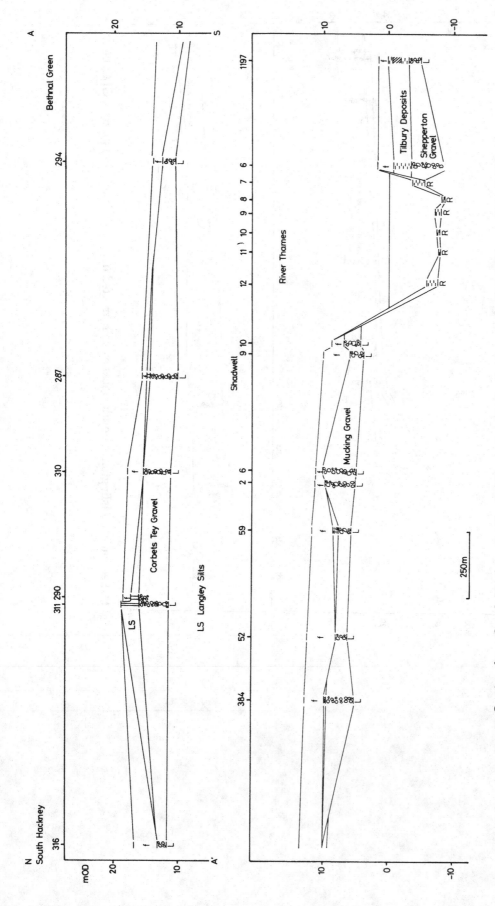

Fig. 17. Section from South Hackney to Shadwell, constructed using borehole data (sources GLC, BGS).

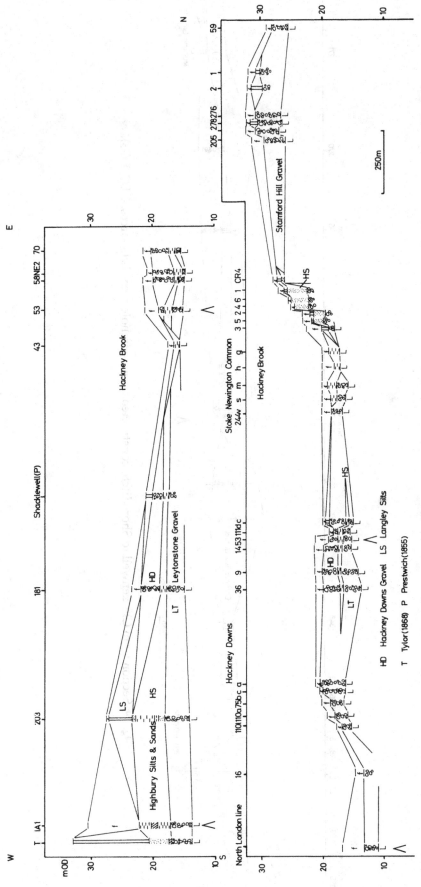

Fig. 18. Sections in the Hackney Downs area. The upper section is orientated W–E and the lower N–S. The sections were constructed using borehole and section data. The borehole data were obtained from BGS and GLC sources, with the addition of data collected by the author in the area north of Stoke Newington Common and at Hackney Downs.

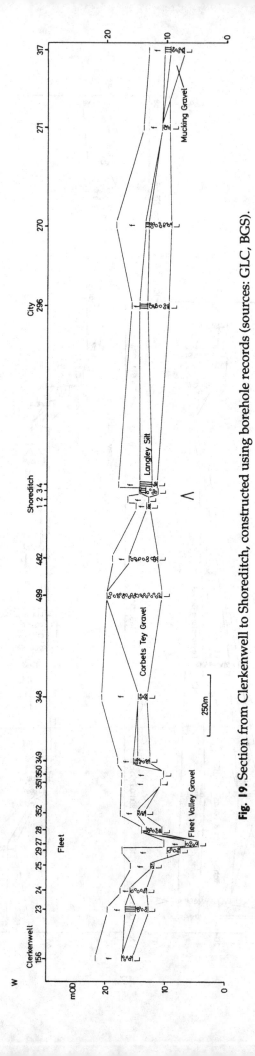

Fig. 19. Section from Clerkenwell to Shoreditch, constructed using borehole records (sources: GLC, BGS).

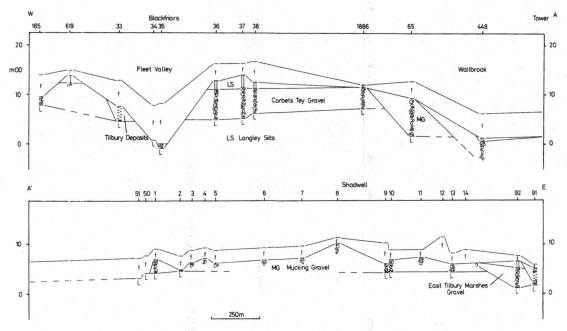

Fig. 20. Section from Blackfriars to Shadwell, constructed using borehole records (sources: GLC, BGS).

Remnants of a possible Lea Valley–equivalent of the Corbets Tey Gravel can be identified on the basis of upstream gradient projection. This Lea Valley unit will be referred to here as the Stamford Hill Gravel, from the gravel spread at Stamford Hill (stratotype: TQ 340877: 32 m). A temporary section on Clapton Common revealed over 1.0 m of sorted gravel and sand overlain by brown clayey gravel and silty sand 2.8 m thick. According to borehole evidence in this patch (Fig. 18), the gravels and sands reach a maximum thickness of 8 m, confirming previous descriptions by W. G. Smith (1894) and Wymer (1968). A palaeocurrent measurement from the tabular cross-bedded sand gave a flow towards 160°, and pebble counts are given in Appendix 2.

Upstream projection of a gradient of 1.6–1.75 m km^{-1} from Wanstead/Fairlop through the outcrops at Leytonstone, Wanstead Station and Whipps Cross, passes through that at Stamford Hill and up the Lea to intersect the patches mapped as Boyn Hill equivalents at Broomfield House (TQ 310925: 40 m) and Bush Farm (TQ 324956: 44 m) (Fig. 22).

2.6 Mucking Gravel

Following their investigations in the Lower Thames Valley, Hinton & Kennard (1901, 1907) recognised a distinct Middle Terrace, intermediate in height between their High (Boyn Hill or Orsett Heath) Terrace and their Third Terrace, the latter immediately above the modern floodplain. During the Geological Survey mapping of the area, the Middle Terrace was equated with the Taplow of the Middle Thames by Dewey & Bromehead (1921), Dewey *et al.* (1924), Bromehead (1925) and Dines & Edmunds (1925). The term Taplow Terrace was

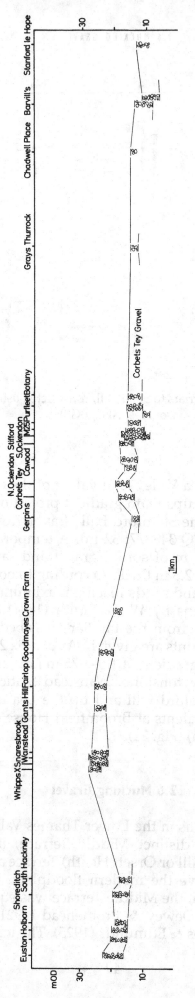

Fig. 21. Long profile section showing the deposits correlated with the Corbets Tey Gravel Member.

2.6 MUCKING GRAVEL

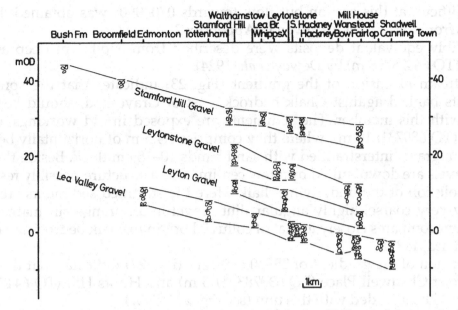

Fig. 22. Long profile section showing the deposits in the Lea Valley and their proposed correlations.

originally proposed by Bromehead (1912) for the deposits exposed at Taplow Station in the Middle Thames Valley, near Maidenhead, and Dewey & Bromehead (1915) applied the term Taplow Gravel following mapping of the Windsor–Chertsey district.

In general no attempt was made to separate the terrace surface from the underlying sediments, with the result that both gravels, and the 'brickearth' that overlies them in places, have been grouped as Taplow or Middle Terrace. As in the Middle Thames (Gibbard, 1985), there is a substantial time gap between deposition of the 'brickearth' and the underlying gravels. The two members are therefore described separately here (Langley Silt Complex, section 2.10).

The recognition of the Corbets Tey Member in the Lower Thames beneath ground mapped as Taplow by the Geological Survey (section 2.5), implies that much of the lower terrace areas shown as Floodplain Terrace should generally be underlain by Taplow–equivalent Mucking Gravel. This indeed seems to be correct (as shown below). The term Mucking Gravel was first proposed by Bridgland (1983a) to refer to a distinct unit of deposits exposed in the eastern Lower Thames Valley. This unit has subsequently been correlated with the Middle Thames' Taplow Gravel on the basis of gradient and stratigraphical position (Gibbard, 1985; Gibbard, Whiteman & Bridgland, 1988; Bridgland, 1988a).

The type section comprises a large working gravel pit, south of Mucking church (TQ 689815: 8 m). Here a maximum thickness of 5 m of stratified gravel and sand was exposed resting on bedrock. The medium gravels in a sand matrix are disposed into 50 cm thick horizontally bedded units separated in places by sand bands or shallow sand channel fillings 15–20 cm thick. The basal zone of the gravel, immediately above the bedrock, includes some boulder-sized clasts up to 2 m in diameter. They comprise mostly sarsen, but some are of flint. The absence of cross-bedded units makes palaeocurrent measurement very difficult at this section but flow towards 070–090° was obtained. Pebble counts from the gravels are given in Appendix 2.

Possibly equivalent deposits were described from a pit 3 m deep at Low Street (TQ 664787: 8 m) by Dewey *et al.* (1924).

Upstream extension of the gradient (Fig. 23) indicates that the spread of deposits banked against Chalk bedrock, east of Gravesend, should be correlated with this member. The sediments are exposed in old workings at East Court (TQ 689731: 10 m), where they comprise *c.*4.5 m of horizontally bedded, medium gravel interstratified with sand bands 10–15 cm thick. Beside the road the gravels are downfaulted by a few centimetres: a structure possibly resulting from solution of the underlying chalk. Here the stratified sediments are overlain by very coarse, highly angular flint gravel in an orange silt matrix. This sediment compares closely to that of colluvial origin in chalk bedrock areas (e.g. cf. Gibbard, 1985).

Extension of the gradient of 25–30 cm km^{-1} (Fig. 23) indicates that the small outliers at Chadwell Place (TQ 638783: 11.5 m) and Hunts Hill (TQ 644782: 10 m) should be included with this unit (see Fig. 29 below).

Altitudinally equivalent deposits are next encountered in the Crayford–Erith area, where they underlie the substantial spread of fine sediments collectively called the 'Crayford Brickearth'. At present there is almost no exposure in the area, it having been considerably redeveloped. However, from the early nineteenth century the area was worked for 'brickearth' and was the subject of many geological investigations (see Wymer, 1968, for summary). The most important geological works on the area were by Chandler (1914) and Kennard (1944). The 'brickearth' will be discussed below (section 2.7e); however, over the whole area this material rests on gravels and sands banked against Chalk and Thanet Sand. The gravels reach a maximum observed thickness of about 5 m, and occur at 9 m OD. Elephant teeth and rolled artefacts have been recovered from this gravel (Kennard, 1944; Wymer, 1968). According to contemporary accounts (summarised by Kennard, 1944), the gravel was continuous with that to the east at Slades Green, where the 'brickearth' wedged out. Modern examination of the area supports earlier reports, notably illustrated by Chandler (1914), that the Slade Green gravels probably represent a younger spread, abutting the older gravel beneath the 'brickearth' (see Fig. 33, section 2.7e). This is based on the lower elevation of the former, i.e. 5 m (surface) to 0 m (base) (Kennard, 1944). Boreholes sunk by the author beside the railway at TQ 518772 (7.6 m) proved that gravel occurs here, but it was not penetrated.

An exactly similar situation occurs in the Grays–West Thurrock area, where a closely analogous spread of 'brickearth' rests on gravel up to 3 m thick (Morris, 1836; Tylor, 1869; Dewey *et al.*, 1924) at an appropriate height for possible correlation. According to these authors and Hinton & Kennard (1901) and Dewey *et al.* (1924), the gravel is very variable in thickness across the sites, both north–south and east–west. Reports of shells in the basal gravels (e.g. Morris, 1836; Dawkins, 1867) suggest, however, that part if not all the gravels beneath the 'brickearth' may not relate to the Mucking Gravel, but to an intermediate (younger) depositional event represented by the Spring Gardens Gravel (Gibbard, 1985). Unfortunately, in recent decades sections in the famous Globe or Celcon Pit in Grays Thurrock (also called Little Thurrock: TQ 624783: section 2.7b) have shown very little of the sequence (West, 1969; Hollin, 1977; Wymer,

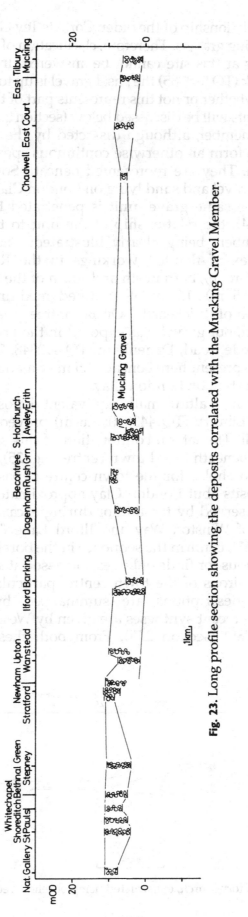

Fig. 23. Long profile section showing the deposits correlated with the Mucking Gravel Member.

1957, 1968), except the relationship of the older, Corbets Tey Gravel to the younger 'brickearth' and underlying gravels. Therefore, clarification of questions regarding the precise relationships at this site cannot be answered. In the Lion Tramway Cutting in West Thurrock (TQ 598785) the basal gravel is up to 1 m thick resting on the Chalk. It is unclear whether or not this represents part of the same spread. The geology of these sediments will be discussed below (section 2.7c).

The gravels of this member, although dissected by the rivers Ingrebourne, Beam, Roding and Lea, form an otherwise continuous spead from Aveley into eastern central London. They are represented beneath South Hornchurch by 3 m of fine to medium gravel and sand lying on London Clay in old excavations (TQ 528843: 9.0 m). The same gravel unit is penetrated by boreholes in the immediate area (Fig. 24), the relationship of the unit to the Shepperton and Tilbury Alluvium Members being clearly illustrated. The deposits are next exposed in the complex of flooded workings in the Rush Green area of Dagenham (Eastbrook Grove), both north and south of the District Line. At the former (TQ 507858: 10.5 m), 1.7 m of stratified medium gravel and sand occurred beneath 1.25 m of 'brickearth'. On Becontree Heath, no sedimentary structures could be seen, but gravel was exposed in the pond side in 1983. An isolated borehole in Reede Road, Dagenham (TQ 492848: 7.9 m) confirms that 6.6 m of gravels are also present here beneath 1.5 m of made ground and brown silty pebbly sand, and resting on London Clay.

In the Barking–Ilford area, altitudinally equivalent deposits are again found. A borehole at Barking Library (TQ 443840: 7.6 m) proved 4 m of gravel and sand beneath 3 m of fill. The lateral continuation of this spread underlies the interfluve from here to beneath Ilford town centre (Fig. 25), where the deposits were encountered in boreholes for the town centre bypass, Winston Way. In these sections the deposits abut London Clay approximately beneath the railway line. Exposures observed by the author during excavations for the roadwork, at the junction of Winston Way and Ilford Lane (TQ 437862) and on Winston Way (TQ 439863), confirm the sequence in the borehole cross-sections.

The Ilford area is famous for finds of bones and associated fossils from brickyard sections in the environs of the town centre, particularly during the last century. The earliest descriptions are summarised by Whitaker (1889, pp. 410–15), while more recent syntheses are given by West, Lambert & Sparks (1964) and Wymer (1985) (section 2.7i). From both these and the original

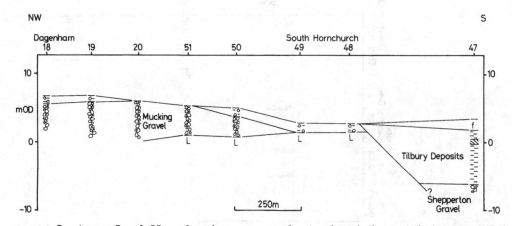

Fig. 24. Section at South Hornchurch, constructed using borehole records (source: BGS).

2.6 MUCKING GRAVEL

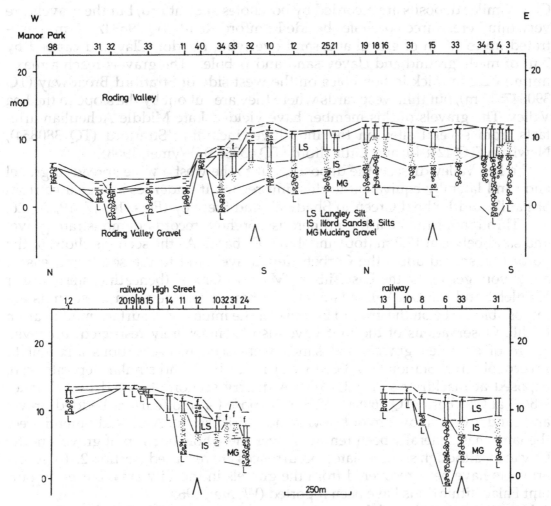

Fig. 25. Sections beneath Ilford town centre. The upper section is orientated W–E and the lower sections are N–S. The sections were constructed using borehole and section data. The borehole data were obtained from BGS, GLC and ERCU sources, with the addition of data collected by the author.

accounts to which they refer, it is certain that stratified gravels and sands underlie the fossiliferous sand and fine sediments ('brickearth') over most of the area. The latter are described below (section 2.10), but on the basis of their height and disposition, there can be little doubt that the gravels belong to this member.

The southern part of the Lea–Roding interfluve is also underlain by the Mucking Member, its outcrop extending from Aldersbrook to Newham. This conclusion is contrary to the outcrops shown on the Geological Survey map sheet 257. The Geological Survey mapped the deposits in this area as 'Taplow', while the lower part was mapped as 'Floodplain', although they admit 'their boundary lines, as shown upon the map, are to some extent conjectural' (Dines & Edmunds, 1925). At Wanstead Flats, where a well-developed terrace surface can be seen, a poor exposure in the bank of the pond (TQ 417862: 13.7 m) showed gravel and sand in a clay matrix. Further south the terrace surface is gently degraded, such that at Barking Road, Upton Park (TQ 4198834: 7 m) 0.9 m of fill overlie 5 m of stratified gravel and sand, which in turn rests on London

Clay. Similar deposits are recorded by boreholes at Stratford, but the gravels are very thin here. Three boreholes beside Romford Road (TQ 396847: 12 m) penetrated up to 1.5 m of gravel and sand overlying London Clay and capped by 2 m of made ground and clayey sand and pebbles. The gravels reach a maximum of 2.2 m thick in boreholes on the west side of Stratford Broadway (TQ 390847: 11 m), but thin westwards where they are cut out by the slope to the Lea Valley. The gravels of this member have yielded Late Middle Acheulian artefacts from several sites on the interfluve, including Stratford (TQ 380854), Newham (TQ 424840) and Little Ilford (TQ 433853) (Wymer, 1968).

The Lea Valley dissects the deposits, but to the west a wide spread of gravel and sand laterally equivalent to this unit recurs. It underlies a large area from St Paul's and Bethnal Green to Shadwell and Stepney (Figs. 16, 17, 19, 20, 26, 34). Throughout the area the numerous borehole records demonstrate gravel and sand between 12.2 m (top) and 4 m OD (base). As the sections show, to the north this spread abuts the Corbets Tey Gravel, and to the south and east it abuts younger units: the East Tilbury Marshes Gravel (beneath eastern Tower Hamlets) and the Lea Valley Gravel. The boundaries between these units are impossible to see on the ground because of the intensity of surface modification by fill. Observations of the unit have also been severely restricted; however, 2.5 m of stratified gravels and sands were seen in excavations adjacent to Liverpool Street Station (TQ 333816: $c.11$ m) in 1990, and similar deposits were exposed at Blackfriar's (TQ 317811). A quarry section observed by Whitaker (1889) at Old Ford (TQ 369841: 12.5 m) exposed 4 m of current bedded gravel and sand, overlain by 2 m of brown sandy 'brickearth'. As already mentioned, the latter has generally been removed over the whole outcrop of gravels in the City area, although some isolated occurrences are recorded (section 2.10). A few artefacts have been recovered from the gravels in the City area, but no important Palaeolithic finds have been reported (Wymer, 1968).

Upstream extension of the gradient demonstrates that the Mucking Member passes directly into the Taplow Gravel west of the Fleet valley, the closest Taplow outcrop occurring beneath the National Gallery (Gibbard, 1985). There can be no doubt therefore that the two are lateral equivalents, as suggested by Gibbard (1985) and Bridgland (1988a). The downstream gradient throughout the Lower Thames region is 25–30 cm km^{-1} (Fig. 23. The pebble lithological composition comprises 90–94% flint (58–75% angular flint, 15–35% rounded flint), 2–9% vein quartz, 0–2.5% quartzite and 0–3.5% Greensand chert (Appendix 2).

In the Lea Valley a unit that passes laterally into the Mucking Gravel can be identified. This unit is present in boreholes for the Hackney–M11 link road in Leytonstone (Fig. 14), where it forms a discrete unit between 21 m (top) and 13.2 m OD (base). Similarly, in the Finsbury–Roding Valley section (Fig.15) a unit at 23m (top) - 15m OD (base) is present beneath Walthamstow. Plotting of these units strongly suggests that they represent patches of a single unit and indicates that this would be confluent with the Thames' Mucking Gravel at Wanstead Flats. The gradient on this unit is 1.5 cm km^{-1}. Since it is best developed beneath Leytonstone town centre, the term Leytonstone Gravel is proposed for this unit (Fig. 22).

Detailed investigation in the Hackney Downs–Arsenal area has already been discussed above (section 2.5). North–south and east–west transects across the

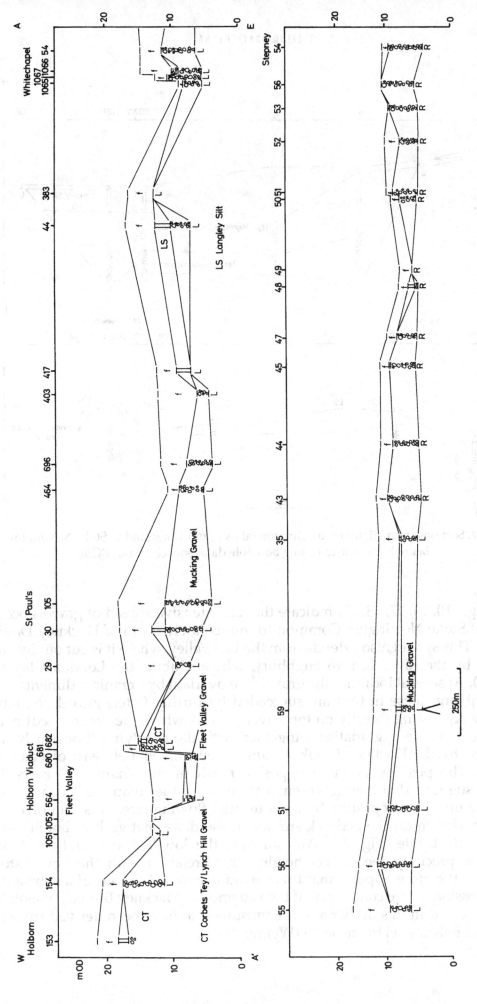

Fig. 26. Section along the Central Line from Holborn to Stepney, constructed using borehole data (sources: GLC, BGS).

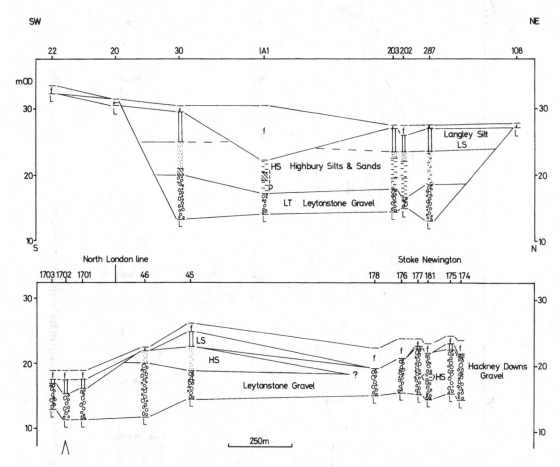

Fig. 27. Sections from Highbury to Islington (above) and Kingsland to Stoke Newington (below), constructed using borehole data (sources: GLC, BGS).

area (Figs. 10, 16, 17, 18, 27) indicate that an extensive spread of gravels occurs south of Stoke Newington Common to immediately north of Hackney Downs Station. This spread also extends from the Lea Valley, where it is cut out by later erosion by the River Lea, to Highbury, where it abuts the London Clay hill (Fig. 27). At several localities the gravels are overlain by organic sediments (section 2.7j), and these in turn are succeeded by further (later) gravel, the latter presumably resting directly on the lower gravels where the organic sediments are absent. These aggradations, together with the overlying 'brickearth', are dissected by the Hackney Brook stream valley immediately east of Hackney Downs. The position of these deposits, north of the Thames' Corbets Tey Gravel, suggests that they represent a tributary, rather than the Thames itself. That the unit was deposited by a southward flowing stream is reinforced by contours drawn on the bedrock surface in the dense network of boreholes at Nightingale Estate (Fig. 28). Altitudinally, the lower gravel and sand unit occurs at precisely the correct height for correlation with the Leytonstone Gravel. It therefore appears that this spread originated as part of a substantial bend, possibly at the confluence of the Lea and the Hackney Brook. Palaeolithic artefacts occur in this outlier and are presumed to have been derived from the Stamford Hill Gravel to the north (Wymer, 1968).

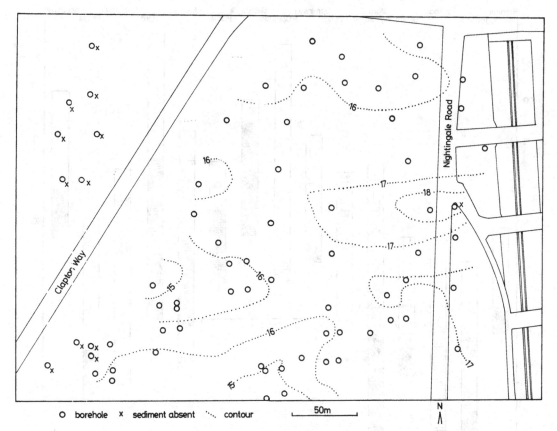

Fig. 28. Map showing the elevation of the surface of the Leytonstone Gravel (immediately beneath the organic sediments) beneath Nightingale Estate, Hackney Downs (source: BGS).

2.7 Aveley/West Thurrock/Crayford Silts and Sands

In this section the sequences at a series of key sites are described and discussed. Many of the sites are either no longer exposed or only partially available for investigation. For these the previous descriptions are assessed and, whenever possible, interpreted in the light of modern observations. Other sites found or re-examined during this study are also included. The descriptions that follow are restricted to the lithostratigraphy of the localities. The sequences are summarised in Fig. 29. The biostratigraphy is discussed separately in Chapter 6.

(a) Northfleet

At Northfleet in the small south bank Thames' tributary valley of the Ebbsfleet, sections in a chalk quarry have yielded abundant Palaeolithic artefacts: in particular the Baker's Hole industry, originally described by Smith (1911). The sequence, investigated in detail during the 1930s by Burchell (1933, 1935, 1936a, 1936b, 1954, 1957), comprises a series of stratified gravels, silts and chalk solifluction ('coombe rock'). The history of the sites is summarised by Wymer (1968, pp.354–9), and a description of the 1960s British Museum excavations is presented by Kerney & Sieveking (1977). Since the sites were not investigated in detail during this study, only the main observations and conclusions are presented.

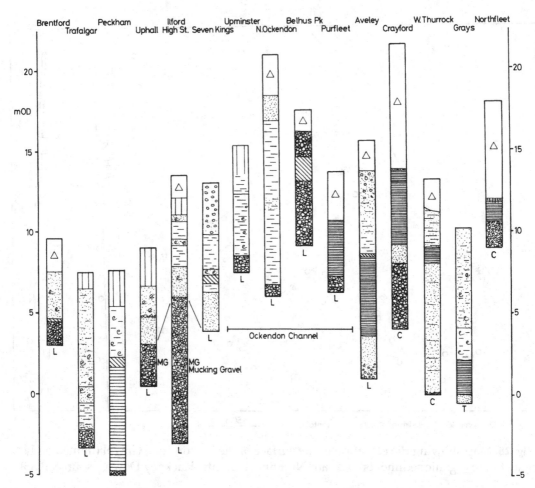

Fig. 29. Summary of sediment sequences of the main localities correlated with the Aveley Silts and Sands Member.

From the descriptions it seems that there was a series of sites along the western valley side. The first, Baker's Hole (TQ 618736) has now been completely removed by quarrying, but according to Wymer (1968) and Dewey (1932) the sequence comprised 1.5–3 m of chalk debris, Tertiary pebbles and angular flints in a putty-like chalk-rich clay matrix ('rubble chalk' or 'coombe rock') resting on Chalk bedrock at 12.2 m OD. The 'coombe rock' was overlain by c.2.5 m current bedded gravel and sand, the base of this unit being channelled into the underlying sediment. This in turn was overlain by 3–6 m of 'brickearth' that included a molluscan fauna typical of cold climates. Within the 'brickearth' a weathered horizon (presumably a palaeosol) was overlain by 1.25 m thick silt bed containing temperate climate Mollusca. Prolific Levalloisian artefacts have been recovered both from beneath and in the 'coombe rock' as well as numerous remains of mammoth, rhinoceros, horse and deer. Burchell's additional details of this sequence are given by Zeuner (1959, p. 164).

Investigation of the palaeosol in the 'brickearth' (on the 'Middle Loam') and the associated sediments by Kemp (1991) has demonstrated that the soil represents a true stable land surface. This surface developed in the loam, the latter originating by slope wash. The micromorphology of the soil indicates that wetter conditions arose after a period, presumably associated with local watertable rise. This was followed by the re-establishment of the flowing water conditions at the site that led to deposition of the overlying shell-bearing silt bed. The sequence was then truncated by the 'Upper Loam' or 'brickearth' that comprised a 'laminated sand-loam' at the base (S.C.A. Holmes, unpublished).

The so-called British Museum site (TQ 611742) is about 300 m north of Baker's Hole, where the sediments were still exposed during this study. Here again massive 'coombe rock' resting on Chalk bedrock is overlain by sorted deposits, the base of which is at 9 m OD (Fig. 29). The latter consists of a lower gravel and sand and an upper silt 'brickearth'. The sequence at this site can be equated with that at Baker's Hole by the freshwater mollusc-bearing silt bed. The silts are weathered at the top, thought by Kerney & Sieveking (1977) to indicate an hiatus. Overlying the silts are up to 6 m of sandy colluvial periglacial deposits, interbedded with chalk-rich 'solifluction' horizons.

Kerney & Sieveking conclude therefore that at least two cold stage accumulations are represented here, during which substantial quantities of colluvial material, 'rubble chalk' and 'brickearth' were deposited on the valley side. The 'brickearth' has also been shown to include a substantial loess component, both before and after the temperate interval. These cold stage deposits are separated by a temperate event, during which freshwater sediments and, presumably following a period of local drying out, weathering occurred. The molluscan assemblage in the temperate sediments has been correlated with the Ipswichian Stage (Ip IV) by Kerney & Sieveking – an age that implies the underlying cold stage sediments are immediately pre-Ipswichian (Wolstonian) and the overlying deposits are post-Ipswichian, i.e. Devensian. The pre-Ipswichian age of the lower 'brickearth' sediments has been confirmed independently by Wintle (1982) and Parks & Rendell (1992).

(b) Grays Thurrock (Little Thurrock and Orsett Road)

The mapped spread of 'brickearth' almost 1.5 km long from Grays station to Little Thurrock has been almost entirely dug away. Excavations during the last century, mostly for brick-making material, yielded rich finds of fossil, particularly mammalian remains, but also Mollusca, plant remains and Palaeolithic artefacts. The earliest report is by Morris (1836), but descriptions of the sequence have regularly appeared since (e.g. Dawkins, 1867; Tylor, 1869; Whitaker, 1889; Hinton & Kennard, 1901, 1910; Kennard, 1904, 1916; Dewey *et al.*, 1924; King & Oakley, 1936; Wymer, 1957, 1968, 1977; Snelling, 1964; West, 1969; Hollin, 1977). The main sections were those in Little Thurrock, Orsett Road (TQ 620783) and the Globe Pit or Celcon Works (TQ 624783), the latter being still in existence, although there is virtually no exposure at the site.

The sequence reconstructed from the reports cited above is summarised in Figs. 29 and 30. The following elements can be recognised:

1 A basal gravel and sand resting on Chalk or basal Tertiary bedrock at about −1 m OD. The gravel is described by Morris (1836, p. 262) as 'a bed composed of rounded and angular chalk, and iron, flints, quartz, sandstone, indurated claystone, nodules of white chalk and calcareous marl, fine sand and comminuted fragments of shells, exposed about 4 ft (1.2 m).' Dewey *et al.* (1924, p. 105) report that the gravel reached 'up to 10 ft (*c*.3 m) in thickness'. They note that the basal gravel and sand fills a broadly

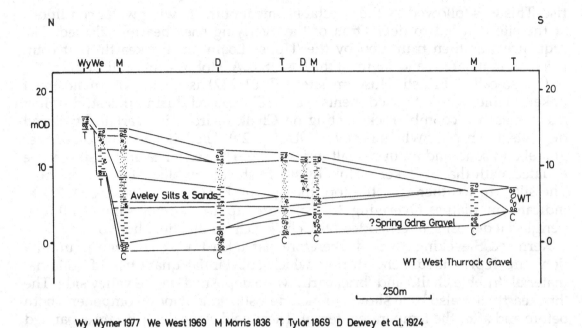

Fig. 30. Section from the Grays Thurrock (Little Thurrock) quarry, based on the sources quoted.

west–east trending depression in the bedrock, abutting a steeply-rising bedrock Chalk and Thanet Sand cliff-like wall on the north side of the spread. This slope 'was covered by sands with a few flints, apparently soliflucted from the Thanet Beds, and then sandy gravel running into the basal gravel' (West, 1969, p. 278).

According to Wymer (1977, p. 310), Conway found Clactonian artefacts both in and on the surface of this gravel. Wymer confirms West's observations that a sheet of soliflucted gravel derived from his 15 m OD gravel (Corbets Tey Gravel: see above), 'merges into or possibly covers the basal gravel' and implies that this presumably carried with it the Clactonian material from the higher gravel deposit. It is not clear whether the Clactonian material that was also recovered from the overlying 'brickearth' has a similar origin. Contrary to the statement by Hollin (1977, pp. 36–7) the shells present in the basal gravels have not been identified. Mammalian remains have also been recovered from this deposit in a rolled condition (Tylor, 1869). The stratigraphical position of this unit suggests that it might be the lateral equivalent of the Spring Gardens Gravel of the Trafalgar Square area (Gibbard, 1985) rather than a higher member.

2 Resting conformably on the gravels is a unit of laminated clays and sand: the 'brickearth' of most authors. This unit was originally described by Morris (1836, p. 63) as 'layers of the loam and sandy clay (that) appear to be made up of small flat fragments of shells, arranged parallel to each layer, the laminae varying from eighteen to twenty in number in the thickness of an inch [2.05 cm]'. West (1969, p. 278) reports that this sediment is grey to brown sandy silt and sand with irregular bedding at the northern end of the Globe Pit. Morris continues: 'shells are irregularly distributed: some genera are more abundant in one spot than another'. He also records the occurrence of 'lignite' beds of macroscopic plant material variable in

their position. From the descriptions it is not certain precisely which fossils came from which sedimentary horizon. There is general agreement that sandy bands in or adjacent to the laminated clay unit contained abundant, particularly calcareous, fossils.

The abundant fossil finds from these deposits are listed by Hinton & Kennard (1901). The Mollusca included *Anodonta, Unio, Corbicula, Hydrobia, Valvata, Planorbis, Ancylus, Helix, Bithynia, Pupilla*, etc. (Morris, 1836; Hinton & Kennard, 1901; Kennard, 1916; Hollin, 1977). The vertebrate remains included 'an almost complete skeleton of the elephant' (Morris, 1836) together with a range of other taxa. Bird, reptile and freshwater fish fossils are also well represented, while ostracod species include *Illocypris gibba, Herpetocypris bradyi* and notably *Cyprideus torosa*. Pollen analyses of isolated samples from the Globe Pit by West (1969) and Hollin (1977) have also been undertaken (Chapter 6).

3 Overlying the laminated silts in the Globe Pit is a unit of current bedded sand up to 20 feet (6.1 m) thick, the base comprising a pebbly sand bed (1.25 m thick) (Tylor, 1869, p. 83). This bed is described as being 'crowded with shells' (Whitaker, 1889, p. 420), particularly *Corbicula*. The current bedded sand was clearly laterally variable in thickness and generally unfossiliferous, possibly as a consequence of decalcification. According to Whitaker (1889, p. 420) the palaeoflow in the lower part of the unit was 'eastward, down the valley'. The upper part of the unit seems to have been horizontally stratified 'with wavy and transverse layers' (Morris, 1836, p. 263). At Orsett Road this sand unit is missing, the laminated clays continuing up to within 3 m of the surface (Hinton & Kennard, 1901).

On the south side the deposits are truncated by coarse gravel and sand to the south, which appear to represent the lateral continuation of the West Thurrock Member (section 2.8).

Attempts during this study to examine the sequence at the exposure at Grays failed. However, a borehole (Grays A) was put down immediately adjacent to the pavement and allotments on the roadside at TQ 623781 (12 m OD) where *in situ* material was recovered:

0–1.80 m	Made ground.
1.80–2.17	Dark brown silty sand (?buried soil).
2.17–3.60	Light brown to yellow medium to coarse sand to silty sand with some shell fragments and race.
3.60–5.05	Alternating medium to coarse yellow sand with occasional stones, and light grey silt bands (>2 cm thick). Light grey silty clay at 4.26–4.28 m. Silts become more clay rich with depth and include shells.
5.05–6.15	Pebbly sand with many shells. Includes light grey silt 1 cm thick at 5.27 m. Sand becomes very chalk rich and more silty with depth. Below 5.46 m comprises silty sand with shells and chalk fragments.
6.15–8.03	Green-grey silt with occasional shells and chalk fragments. Becomes interbedded fine sand and silty clay (1 cm thick) at 6.55 m, the individual beds narrowing with depth to 1–2 mm.

8.03–8.06	Coarse sand with a few shell fragments.
8.06–8.78	Laminated grey silty clay and fine sand, the latter becoming coarser and yellow in colour with depth.
8.78–8.85	Grey coarse to medium sand with shell fragments.
8.85–9.62	Laminated grey brown silty clay (1–2 cm) and medium sand (1 cm), the latter thickening with depth to 5 cm.
9.62–9.70	Coarse to medium yellow sand.
9.70–9.73	Mottled grey-brown silty clay.
9.73–10.20	Interbedded grey-brown silty clay and fine sand bands.
10.20–10.55	Oxidised medium sand with some chalk clasts and narrow brown silt bands (>3 cm).
10.55–10.75	Coarse oxidised sand with many small pebbles and chalk clasts.
10.75–	Thanet Sand (borehole stopped).

From this record it appears that all three subdivisions of the sequence, recognised by earlier authors, have been recovered. Unit i, the basal sand and gravel, seems to be very thin in the borehole – only 20 cm. Unit ii, the laminated clays and sands, are particularly well developed; although no vertebrate material was found, shells were quite common. The current bedded sands of unit iii may be represented by sediment very similar to that beneath, i.e. interbedded sand and silts. However, the pebbly sand layer at 5.05 m may be the equivalent of such a stratum in the Globe Pit mentioned by earlier workers (e.g. Morris, 1836, p. 262). A series of samples from this borehole has been pollen analysed (Chapter 6).

(c) West Thurrock

Like Grays, the West Thurrock 'brickearth' spread is aligned along the foot of the Purfleet chalk ridge and extends 2.5 km from immediately east of the M25 motorway Dartford Tunnel approach to South Stifford. It has also been exploited like the Grays area, and access to the large chalk quarries have necessitated excavations through it, both in the past and more recently. The main exposure of this sequence is the Lion Cement Works Tramway (railway) Cutting (TQ 598788). It was originally mentioned by Whitaker (1889, p. 418), and Abbott (1890) found some bones here, but the sequence was only recently described in detail by Carreck (1976) and Hollin (1977). Excavations carried out in the cutting in 1984 by D. R. Bridgland and colleagues (Nature Conservancy Council, now English Nature) and logged by the author confirm earlier descriptions and provided an opportunity to take samples. The sequence can be summarised as follows (Fig. 31).

At the northern end of the cutting (adjacent to Hollin's site WT3: TQ 598782: 17 m) the deposits overlie a shattered, rather irregular chalk surface at 0–1 m OD. This surface rises steeply to the north where, immediately adjacent to the section, the chalk forms a buried, near vertical 'cliff', the full height of the Quaternary deposits. Immediately overlying the chalk is a pebble lag comprising poorly rounded flints. Middle Levalloisian artefacts have been found associated with this gravel by Warren (1942). The chalk surface continues to dip steeply to the south in the sections.

Above the gravel, 8–10 m of yellow sand to silty sand occurs. Throughout, the sands are finely horizontally stratified or laminated, the lamination being

2.7 AVELEY/WEST THURROCK/CRAYFORD SILTS AND SANDS

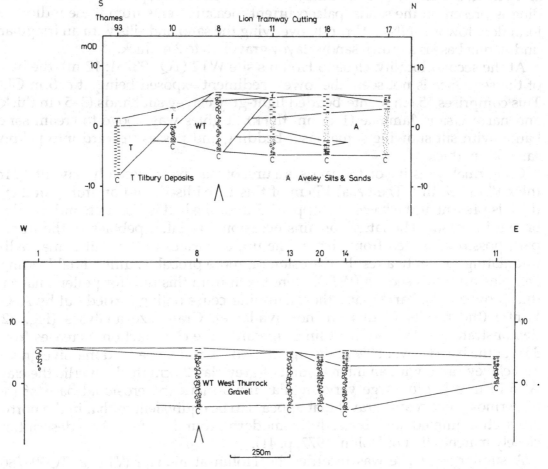

Fig. 31. Sections from West Thurrock. The upper section is orientated N–S parallel to the Lion Tramway cutting and the lower is aligned W–E. The sections were constructed using borehole and section data. The borehole data were obtained from the BGS, with the addition of data collected by the author.

formed by alternation of fine and medium sand layers 1–2 mm in thickness. Some of the coarser sand layers include granules or scattered small pebbles, and above 6 m the sediments become coarser and more crudely bedded. In the basal 2 m, palaeocurrent measurements indicated flow towards the south. On the northern side of the exposure, 'rubble' chalk tongues up to 50 cm in thickness interfinger with the sands. These units thicken towards the buried 'cliff' face on which they abut, but die out 4–5 m to the south. They have erosional bases on the sands beneath, and small sand inclusions, apparently incorporated during their emplacement, occur in the 'rubbly' material. These units appear to represent mass flows. Also interbedded and interfingering with the sands, on the south side of the exposure, are lenses of orange to brown sandy or clayey silt 0–40 cm thick. These units thicken southwards and rest conformably on the sands beneath.

The sands are overlain by 1 m of sandy silt. This unit is orange-brown in colour, except in its upper 20 cm which are grey (a sample from here contained no countable pollen and spores). The upper surface of this sediment is eroded and above it occurs a 2.2 m sequence of interstratified coarse to medium sand (2–10 cm thick) and brown silt bands (1–3 cm thick). Above 10 m some type B ripple cross-bedding is present in the sands; palaeocurrent measurements from these indicate a local flow towards NNE. Abruptly overlying the sand and silts with an irregular, undulating base is a brown sandy clayey gravel up to 2 m thick.

At the second locality, close to Hollin's site WT2 (TQ 598781: 12 m), the base of the sequence is not seen, the lowest sediment exposed being at c. 5 m OD. This comprises 55 cm of interbedded light-grey clayey silt bands (3–5 cm thick) and narrow sand laminae (1–3 mm thick). It gives way above to cream sand bands with silt showing some flaser bedding and passes upward into pebbly sand 30 cm thick.

Conformably resting on the sands is a unit of massive grey sandy clay silt 2.9 m thick ('brickearth'). The basal 15 cm of this is oxidised, and an iron pan 2 cm thick is present at the base. The upper 1.7 m of the bed is clay rich and mottled brown in colour. The latter contains occasional isolated pebbles in the upper part, possibly injected from above. The upper surface of the unit is markedly undulating, possibly a result of erosion or, more probably, differential loading. Samples from this section (WTD) were taken from this unit for pollen analysis (but proved to be barren) and thermoluminescence dating carried out by A. G. Wintle (the results of this are not available). Grain-size analyses (Fig. 32) demonstrate that the sediment fines upwards; the clay fraction increases from 35% at the base to over 60% in the upper metre. Crudely bedded massive gravel and clayey sand, with an intercalation of grey clay 25 cm thick, overlie the clay. At 6.15 m a line of large very angular flints mark the erosional base of the uppermost very coarse gravel that appears to be equivalent to that in the northern section immediately beneath the modern ground surface. This description closely matches that of Hollin (1977, p. 41).

A similar sequence was recorded by Hollin at his site WT1 (c. TQ598780: 9.2 m). Here the basal sand is overlain at 3.5 m OD by 2 m of massive grey clay ('brickearth'), above which a pebbly gravel layer marks the base of a second horizontally bedded sand unit, with scattered pebbles (2.1 m thick). Solifluicted gravel caps this sediment. A section close to WT1 was sampled for pollen analysis in 1984 (WTF; section 6.2). A further section which was cleared 25 m north of the bridge on London Road (TQ 59777792: 8 m) showed a completely different sequence of horizontally bedded medium gravel and sand. This deposit appears to be a younger unit, the West Thurrock Gravel Member, which cuts out and abuts the sediment exposed further north in the tramway sections (section 2.8).

Boreholes immediately west of the cutting (Fig. 31) confirm these observations and demonstrate that the deposits rest on a level chalk surface at c. –4 m OD. This is overlain by about 1 m of gravel and sand which underlies the yellow sands.

In summary, as Hollin (1977, pp. 40–1) notes, the lateral relations of units in the sections are not completely certain. However, the sequence seems to comprise a basal pebble bed on the chalk bedrock, overlain by horizontally bedded sands and silts, occasionally showing current indicators towards the south. In most of the cutting this is overlain conformably by a thick clay unit ('brickearth'). The recent sections clearly show that the latter thins and apparently interdigitates with the sands in the northernmost exposure (WT3). It is not clear whether the silt band in this exposure is indeed the continuation of that further

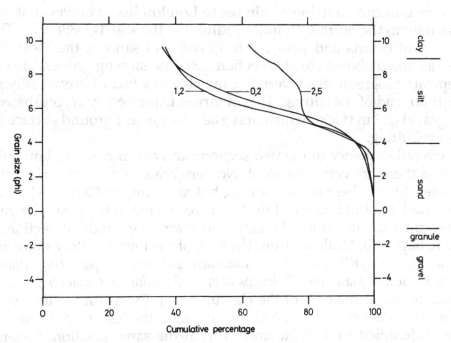

Fig. 32. Particle-size distribution from West Thurrock site WTD.

south or not, since this silt is at 11.3 m OD at the former, i.e. considerably higher than that at WT2.

During this study a second cutting was made for access to the substantial chalk quarry between the A13 and M25, immediately east of Sandy Lane, West Thurrock. A temporary exposure in the cutting, aligned almost parallel with that above, was logged and sampled by the author and D. R. Bridgland (Fig. 1). The section could not be logged in the detail of the tramway since it was open only a short time. However, the sequence was very similar to that already described.

The sediments rest on Chalk that is below the roadway at about 5.6 m OD at the southern end of the section (TQ 590777), but rises northwards to form a steep cliff-like buried surface against which the deposits abut at TQ 589780 (15 m OD). At the northern end of the cutting there is a basal flint pebble lag overlain by orange to yellow fine to medium sands 1.75 m thick. The sands are indistinctly horizontally bedded. Resting conformably on the sands is a unit of brown to grey silty clay ('brickearth'). This was sampled at site WTA (TQ 58987790) where the microstratigraphy was as follows:

	cryoturbated brown clayey sandy gravel
305–335 m	Brown silty sand.
185–305	Orange red mottled light-grey silty clay, becoming silty with narrow sand bands in upper 1.5 m.
60–185	Light-grey silty clay, mottled grey-brown in upper 20 cm, with occasional pebbles and calcareous nodules beneath. Contains pebble layer at 120 cm
40–60	Grey silty clay, purple-grey mottled horizon at 60 cm.
0–40	Interbedded fine sand bands (1–2 cm thick) and light-grey silty clay (1–2 cm), oxidised at base.
	yellow sand

The clay unit can be followed almost to London Road; however, it is cut out immediately to the north by a unit of sand 2 m thick at TQ58987776. This sand thickens southwards and passes into gravel and sand of the West Thurrock Gravel (as noted above: see also section 2.8). The sloping ground also cuts out the deposits progressively. Overlying the clay is a bed of brown silty sand, in the northern end of the cutting. This in turn is truncated by cryoturbated clayey sandy gravel (2.5 m thick), which underlies the present ground surface as in the Tramway Cutting.

The overall sequence in the two sections and accompanying borehole cross-sections is therefore very consistent. No vertebrate or shell material was found at the sites during these investigations, but according to Carreck (1976 and references cited by him) a range of finds was recovered in the past. The vertebrate remains came mainly from the clay and overlying sands at sections in this spread (Chapter 5). Mollusca from the base of the lower, yellow sands included *Helicella itala*, *Pupilla muscoram*, *Lymnaea peregra*, *L. palustris*, *Planorbarius corneus*, *Planorbis planorbis*, *P. leucostoma* and *Bithynia tentaculata*. Additional Mollusca found at the base of the clay unit by J. Evans (in Hollin, 1977, p. 41) include *Anodonta* sp. and *Sphaerium* sp. and the ostracod *Candona* forma 'lacteae' (identified by J. E. Robinson) from the same position. Pollen counts from the clay unit are discussed in Chapter 6.

(d) Sandy Lane, Aveley

The sections in Sandy Lane Quarry, Aveley, occur at TQ 552807 (14 m). Very little has been published on this site, which is surprising since it is one of the most well known in the Thames system. The reason for this is that two fossil elephant skeletons were found, one above the other, at this site in 1964 (Blezard, 1966), and were until recently on display in the Natural History Museum. However, no definitive publication has ever appeared on the site and its important palaeontology. Only three short descriptions are available: Blezard (1966), West (1969) and Hollin (1977). Although sections existed at the site during the present study, they were in a poor state, with large areas rendered inaccessible by dense vegetation. The description here is therefore based largely on the published reports supplemented by the author's observations.

The deposits rest on London Clay at *c*.1 m OD at the deepest part, although the bedrock rises steeply eastwards, such that the deposits are banked against it, as illustrated by Hollin (1977, fig. 5). The lowest sediment comprises 1 m of gravel and sand, overlain by 2.5 m of yellow sand with occasional brown laminae. Organic silts and clays, brown and yellow at the base but dark grey above, rest on the sands. This stratified unit is 4–5 m thick and includes some white silt or clay layers several centimetres thick. Some levels in this sediment are rich in Mollusca, wood and vertebrate remains (the latter described by Stuart, 1974). The *Palaeoloxodon antiquus* skeleton was found at 7.6 m OD in the upper part of the organic silts, whilst the *Mammuthus primigenius* skeleton was 30 cm above (Blezard, 1966, p. 274) in a 60 cm thick bed of detritus mud with much compressed wood that overlies the organic fines. According to Hollin (1977, p. 39), the detritus mud could be traced around the quarry and changed level markedly, from possibly 11 m, to definitely 8.2 m, to below 4.3 m in the centre of the quarry. Conformably overlying, and in places grading up from the detritus mud is a unit of grey-brown to orange-brown massive silty clay ('brickearth').

This deposit is 2.15 m thick and includes carbonate nodules and freshwater shells at the base, found by Cooper (1972). At 9.85 m OD the silty clay grades upwards into 65 cm of grey silty sand, which passes into horizontally bedded yellow sand, with isolated pebbles. This unit reaches a maximum thickness of 3.95 m. Where it abuts the rising London Clay surface to the east, gravel and sand interbeds occur, inclined at 12° W. The sections are capped by brown sandy clay with gravel up to 1.9 m thick.

This sequence described is very limited in lateral extent, on the basis of boreholes for the A13 realignment (Fig. 7), the proposed route of which passes across the western extremity of the quarry. In these the deposits can be traced only 300 m along the section line (NNE–SSW). They are cut out to the north by the London Clay and the East Tilbury Marshes Gravel that occurs at 0–6 m OD in this area, and to the south the bedrock rises steeply, as in the quarry sections. However, the deposits recur in A13 boreholes 316–7, 800 m to the south, where a second 'pocket-like' spread c.200 m in extent is preserved. As at Sandy Lane, this sequence abuts bedrock on both the north and south sides. The sedimentary sequence in this southern patch is also very similar to that at Sandy Lane, although the basal gravel is thicker (9 m) and rests on bedrock at –6.25 m OD. The gravel is overlain by 1.05 m of brown silty sand, which is followed by 2.2 m of mottled grey-brown clay (?'brickearth'). The deposits are sealed by 1.8 m of solifluçted sandy gravel.

It therefore appears that the Aveley sequence represents sediment preserved in the sides of a valley banked against the northern edge of the Purfleet ridge or Thurrock anticline. Later dissection has left these two areas as isolated patches perched on the valley side.

(e) Crayford–Erith

As already discussed, deposits altitudinally equivalent to the Mucking Gravel underlie the substantial spread of fine sediments collectively called the 'Crayford Brickearth' in the Crayford–Erith area. Today there is almost no exposure in this area; however, from the early nineteenth century the ground was worked for 'brickearth' and was regularly investigated (see Whitaker, 1889; Wymer, 1968, for summary). The most important geological works on the area were by Chandler (1914) and the authoritative review by Kennard (1944). As Kennard (1944, fig. 12) shows, the deposits were exposed in a series of workings aligned along the main north–south road from Erith to Crayford. The main localities were Stoneham's Pit (TQ 517758), Rutter's Pit (TQ 514765), Norris' Pit (TQ 514768), Furner's Old Pit (TQ 519768), Furner's New Pit (TQ 520766) and Talbot's Pit (TQ 520763) (grid references according to Wymer, 1968).

The 'brickearth' rested on gravels and sands banked against Chalk and Thanet Sand that rises steeply to the west. These basal gravels reached a maximum observed thickness of about 5 m, and occurred at 9 m OD (section 2.6). According to Spurrell (1885), 'these lowest gravels of this deposit contain rhinoceros, elephant, and other large bones'. In the late nineteenth century, the same author discovered a scatter of Levalloisian implements on the gravel surface, at the base of the 'brickearth' in Stoneham's Pit. This so-called 'working floor' comprised flakes and blades, many of which could be refitted, associated with mammalian remains. Another 'working floor' was later found in Rutter's Pit by Chandler (1916).

The 'brickearth' was a complex of three distinct subunits. The 'lower brickearth' seems to have been the most extensive and yielded most of the fossil material, as well as some Palaeolithic artefacts. Whitaker (1889, pp. 433–9) recorded several sections in these deposits and notes the apparent lateral and vertical variability of the sequence. The base of the lower subunit frequently consisted of yellow to buff sand with shells that passed up into the 'brickearth'. The maximum observed thickness of the sand was 1.8 m, immediately south of Erith. Where it occurred, this sand apparently passed upward into the 'brickearth'. The latter is repeatedly described as 'sandy' (e.g. Kennard, 1944), but no detailed descriptions are given of its sedimentary structure. However, Whitaker (op. cit.) clearly states that the deposit was brown and bedded, or laminated with layers of sand alternating with sandy silt (?'brickearth'), particularly in the lower parts. He even refers to it being 'false-bedded' and 'contorted or wavy', presumably implying current activity and water-release or loading, respectively. Estimates of the thickness of this unit vary considerably, but Whitaker recorded a maximum of 40 feet (12.2 m) at NE Crayford, although thicknesses of up to 18 feet (5.5 m) are regularly mentioned.

Kennard (1944, p. 126) notes that the 'brickearth' contained 'lenticular patches of sand and pebbles', and similar units were described by Whitaker as discontinuous and pebbly sand beds, often no more than 1 m thick. Shell and small mammal remains were apparently concentrated in these sands, but larger mammal bones seem to have been scattered throughout the 'brickearth'. Kennard particularly mentions that the larger bivalves were in life position, i.e. with their valves united.

Overlying the silts and grading up from them was a bed of grey clay, 30 cm thick, particularly in the Northend area (Whitaker, 1889). A laterally impersistent bed of sand of variable thickness (30 cm on average) rested on the clay and 'brickearth' over much of the exposure. This sand was so rich in shells that it was termed the '*Corbicula* Bed' by Kennard (1944). He also states that it 'yielded nearly all the small vertebrates'. The sand bed was 'false-bedded' (Tylor, 1869) and included rolled fragments of 'brickearth', presumably 'rip-up' clasts derived during deposition of the sands. In one area of Rutter's New Pit, over 2.4 m of sand were recorded, but Kennard (1944) was of the opinion that this was 'a sandy phase of the Lower Brickearth'.

Over fifteen species of freshwater and two land species of Mollusca have been recorded by Kennard (1944) from this site.

All the sedimentary units apparently thickened towards the west, against the 'buried bedrock cliff'. The deposits (the so-called 'upper brickearth') overlying those already described thickened markedly in this direction. Tylor (1869) illustrates several sections that show the sediment and Leach (1906) confirms that it was 'thinly bedded and much more clayey than the lower beds.' It apparently included thin, often discontinuous pebble layers and isolated pebbles and passed upward into clayey silt. This sequence was 7.6 m thick at Crayford (Tylor, 1869, p. 94). Adjacent to the bedrock 'cliff', there are repeated descriptions of block-like inclusions of local Woolwich Beds material (see below). In addition, flow type structures and pebble 'stringers' are associated with these beds.

Observations by the author confirm that clay-rich 'upper brickearth' type sediments occur behind houses in Myrtle Close (TQ 514772), where 3.5–4 m of massive orange-buff clayey silt are exposed. Impersistent lenses of flint pebbles, many of typical rounded 'Tertiary-type', and some large isolated chalk clasts 7.5–10 cm in diameter, occur in the silt. In other parts the silt is massive. The consistent impression from these descriptions is that this 'upper brickearth' suite represents material derived predominantly by periglacial colluvial processes from the higher bedrock area, immediately to the west. This conclusion is precisely that already suggested by earlier workers, e.g. Tylor (1869) and Chandler (1914).

The 'upper brickearth' contained few fossils in comparison to the units beneath. Vertebrate remains have been recovered, but almost no Mollusca, except those derived from the local Tertiary rocks. Of the former, *Mammuthus primigenius*, *Ceolodonta antiquitatis* and *Equus ferus* suggest a cold climate regime (section 5.2). Such a regime would equally be expected from the sediments.

Apart from the undisturbed sequence of deposits in this area, there have been repeated observations of large blocks of bedrock and Pleistocene sediments associated in substantial masses adjacent to the 'buried bedrock cliff'. The occurrence and orientation of these structures suggests formation by rotational landslip (e.g. Tylor, 1869; Whitaker, 1889; Chandler, 1914). It is not certain from the descriptions available which sedimentary unit these slip-block masses are associated with.

The uppermost material is a 'dark brown sandy clay with pebbles and broken shells (from the Woolwich Beds), 3–6 feet, resting with an irregular pipey junction on the next' (Whitaker, 1889, p. 436). Chandler (1914, p. 66) describes the sediment in detail, commenting that 'the 'trail' is composed of local materials …derived from the higher ground in the immediate vicinity, over the surface of which it appears to have moved in a semi-fluid stream. Thus it contains Tertiary pebbles and materials from terrace gravels in a paste of clayey sand from the brickearth and Tertiary beds. It rests very unevenly upon the underlying beds, and contorts the brickearth,…the base of the 'trail' is festooned and contorted.' He also notes that 'pebbles frequently have their long axes vertical and this can be explained by supposing the soil to be frozen.'

Boreholes at Perry Street on the south side of the road to Dartford all indicate that much material has been removed and replaced by made ground (1–9 m thick). Beneath this fill, brown silty clay with layers of sand and occasional pebbles, 1.4 m thick, rests on gravel and sand 1–5 m thick. However, possibly thicker *in situ* material is recorded in two holes at TQ 52057607 (11 m) and 52187592 (11 m), which under 1 m of fill proved 2 m of brown clayey sand and gravel upon 3.8 m of 'brown sandy silty clay with layers of light grey silt and sand', in turn resting on gravel and sand (2.6 m thick). The chalk bedrock surface was at 1.4 m OD here. This sequence is confirmed in the borehole section in Fig. 33.

Numerous attempts were made to recover *in situ* sediment in the Crayford–Erith area during this study, but although some was found, little can be added to the descriptions above. For example, beside the railway at Northend (TQ 51807725) the following sequence was found:

	ground surface 7 m OD
0–1.25 m	Made ground.
1.25–1.45	Light brown sandy clay with stones.
1.45–2.40	Yellow-brown silty sand, with narrow bands of clayey sand interbedded with sand and pebbly sand below 2.0 m.

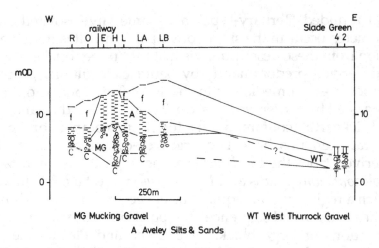

Fig. 33. Section from Erith to Slade Green (Crayford), constructed using borehole data (source: BGS).

2.40–2.90   Sandy fine gravel.
>2.90   Gravel (hole stopped).

This hole appears to have penetrated the basal gravel, the yellow sand and the colluvial material of the 'upper brickearth' unit. In contrast, several holes were put down in the Thames Road–Northend Road area. Of these a hole in the rear garden of Northend Post Office (TQ 51627611) proved:

                ground surface 15 m OD
0–1.50 m   Light brown clayey silty sandy gravel, becoming more clayey with depth.
1.50–1.90   orange sand, clay silt band 1 cm thick at 1.60 m, becoming fine gravel and sand (hole stopped).

Similarly, a borehole on the south side of Crayford Road at TQ 51677600 showed:

                ground surface 17 m OD
0–2.15 m   Made ground.
2.15–2.37   Orange mottled brown clay with a trace of sand.
2.37–2.60   Light grey sand and silty clay interlaminated.
2.60–2.80   Finely bedded sandy silt with small calcareous nodules.
2.80–3.25   Orange brown slightly sandy silty clay with occasional narrow sand bands, becoming grey with depth. Tertiary shell fragments at 3.00–3.10.
3.25–3.70   Brown very dense clay with calcareous nodules (hole stopped).

Both the latter two boreholes suggest that the 'upper brickearth' was penetrated: the former directly onto underlying gravel and sand; the latter apparently did not bottom the 'upper brickearth' unit. These unsuccessful attempts to obtain fossil-bearing sediment mirror a similar attempt by Hollin (1977) and suggest that little undisturbed sediment remains in the area except possibly beneath old roads.

2.7 AVELEY/WEST THURROCK/CRAYFORD SILTS AND SANDS

(f) Purfleet

The sediments at Purfleet are exposed in two large chalk quarries, called Greenlands (TQ 568785) and Bluelands (TQ 570797) on each side of the small lane, North Road. Here the ground surface is at 15 m OD. The deposits were originally described by Snelling (1973/4) and an archaeological excavation was reported by Palmer (1975). Hollin (1977) and Wymer (1985) have both summarised the site sequence.

These deposits were exposed during the present study and appear to fill a channel-like depression in the Chalk, aligned ENE–WSW against the rising bedrock of the Thurrock anticline (Purfleet ridge), the latter visible immediately beneath the ground surface in the southern side of Greenlands quarry. The author's observations of the sequence confirm those of previous authors. The sediments filling the channel are cut into an earlier gravel spread, the Corbets Tey Gravel (section 2.5).

The sediments rest on brecciated chalk bedrock, the surface of which reaches a minimum of 6.25 m OD. Above is a bed of coarse gravel and sand, with some chalk clasts up to 5 cm in diameter and some shell fragments. In Bluelands Pit this unit is 50–100 cm thick, but reaches 1 m in Greenlands (Appendix 2). This passes upwards conformably into 45 cm of orange sandy fine gravel containing unbroken shells. In both exposures laminated clayey silt rests on the sediments beneath with a distinct non-sequence. The laminated bed comprises laminae of sandy silt alternating with silty clay up to 5 mm thick. The basal 55 cm of this unit is grey in colour, but above it is brown. There is considerable detail in the microstructure of the laminae, individual horizons showing water release structures, as well as microchannels 2–3 cm deep filled with coarse sand, fine granules and chalk fragments. The maximum thickness of this unit observed by the author was 2.1 m in Bluelands quarry; however, Hollin (1977, p. 38) records 3.5 m at the same site, the highest laminae occurring at 10.6 m OD or possibly higher (Fig. 29).

In Greenlands quarry the laminated unit is only 1.4 m thick over much of the exposure. Here it is cut out by a channel fill of orange-yellow sand *c*. 40 cm thick, overlain by brown sand full of Mollusca, particularly *Unio* sp. Hollin (1977, p. 38) remarks that 'the molluscs occur in every orientation, suggesting that they are not in position of life, but the bivalves are still paired, suggesting they have not drifted far.' The fill is completed by ?30 cm of grey silt. Sand 70 cm thick, probably equivalent to that in the neighbouring section, is also present in the south faces of Bluelands Pit resting on the laminated bed.

Unconformably overlying the sand and laminated beds alike is a coarse to medium sand matrix supported gravel. Although the gravel appears massive, two fine gravel partings were visible in the south face of Bluelands quarry in 1983. In the adjacent sections in Greenlands quarry, Hollin (1977, p. 38) notes that the gravel tends to alternate with sand in horizontal beds about 20 cm thick. This unit is very rich in subrounded chalk clasts and was likened to 'coombe rock' by Palmer (1975, p. 3). Redeposited carbonate coats clasts of all lithologies, particularly in the lower part of the deposit. The unit is variable in thickness, but reaches about 3 m in Bluelands. In Greenlands, sections 15–20 m south of the lane showed up to 5 m of these horizontally stratified chalk-rich gravels interbedded with sand beds 15–20 cm thick, immediately below the modern ground surface. No current bedding is present, but the gravel bands are cryoturbated. This unit appears to be a continuation of that overlying the channel sediments. A fabric from this unit indicated a dip towards 25–35°. Palmer (1975) interpreted this as indicating possible stream flow from the northeast, but it appears to be a mass flow deposit to judge from its structure and form.

Immediately overlying the gravels and underlying the modern ground surface in the Greenlands section is a brown sandy silt with isolated pebbles, 30–40 cm thick. This colluvial sediment is the lateral equivalent of the bed that caps the Bluelands sections, described as 'brickearth' by previous authors. In Bluelands the deposit is 3 m thick, the upper 1.5 m being massive and clay rich. Beneath it contains a pebble band 20 cm thick (section 2.10).

Artefact finds recovered under controlled conditions reported by Palmer (1975) and discussed by Wymer (1985) indicate that possibly three industries occur here. Clactonian hand axes in rolled condition occur in the basal shelly gravel and the 'middle gravel' or chalk-rich unit, immediately above the laminated bed. In addition, a Levallois flake was found in the uppermost gravel, in the 'brickearth' at Bluelands.

Nearby sections investigated in 1986 at the Esso oil terminal at TQ 560785 (13 m OD) by the author and D. R. Bridgland showed a potentially laterally equivalent sequence to that just described. Here the chalk bedrock surface at *c.* 8 m OD was overlain by 40 cm of fine sandy gravel. This was overlain 60 cm of type B ripple bedded fine sand, which was truncated by a chalk-rich gravel 2.4 m thick. This gravel unit fined upwards and included two beds of parallel cross-bedded sand 20 cm thick. The foresets in the sands indicated a palaeocurrent towards the south-west. Laminated interbedded fine sand and silt 40 cm thick rested conformably on the sands and was capped by a further sand bed 50 cm thick. No fossils were found at this site. The relationship of this unit to the neighbouring Botany Pit exposures (section 2.5) could not be established with certainty.

If this Esso site sequence is indeed the lateral extension of that at Bluelands and Greenlands, it indicates that the channel extended to the Thames close to where the Mar Dyke now enters the river.

(g) Belhus Park, Aveley

Exposures in cuttings for the construction of the M25 motorway in November 1979 were examined by the author and colleagues. These exposures, at a bridge construction site (TQ 575810: 17 m), were also examined by Ward (in Wymer, 1985).

The sequence here comprises 2.7 m of horizontally bedded gravel and sand resting on London Clay. This thickness was not fully exposed but was confirmed by boreholes being undertaken at the site at the time of the visit. It is also confirmed by boreholes for the bridge construction (Fig. 12). The gravel and sand has been correlated with the Corbets Tey Gravel (section 2.5).

Overlying the gravel and sand with a conformable but irregular base is an organic deposit up to 1.5 m thick. The organic sediments are succeeded unconformably by a second unit of horizontally bedded gravel and sand, 2 m thick. The sections are capped by up to 1.2 m of brown pebbly clay.

The organic sediments were sampled in detail for palaeontological analysis (chapter 6) in the western cutting face, where the following sequence was recorded:

	gravel and sand
1.12–1.24 m	Oxidised orange-brown silty clay.
1.04–1.12	Mottled grey-brown silty clay.
0.98–1.04	Light grey silty clay with wood fragments, some large pieces.
0.86–0.98	Light grey clay with wood and plant fragments.
0.57–0.86	Grey silty clay mud with wood fragments.
0.38–0.57	Dark grey silty clay mud with wood and plant fragments.
0.24–0.38	transition to:
0–0.24	Light grey pebbly clay with a few plant fragments.
	gravel and sand

An almost identical sequence, but containing fewer wood remains, was also exposed on the eastern face of the 30 m wide cutting. These deposits were also encountered in the boreholes for the M25 motorway (Fig. 12). Here they occur for just over 1 km, being cut out to the north and south by dissection of the land surface and by old gravel workings. However, the altitudinal and stratigraphical position of the deposits suggests that they are almost certainly the lateral equivalent of those at North Ockenden and Upminster (see below). Similarly, they may also be the lateral equivalent of the channel-fill sediments at Purfleet (see above).

(h) North Ockendon and Upminster

The boreholes for the M25 motorway (Fig. 12), already referred to above, penetrate a substantial series of clays, silts and sands filling a channel-like depression in the area near North Ockendon. These deposits rest on London Clay at a minimum depth of 6 m OD. Resting on the bedrock is a unit of gravel, sand and pebbly sand up to 5.9 m thick. This is overlain by a complex interbedded sequence of sand, pebbly sand and grey silty clay, illustrated in Fig. 29. This sequence is internally very variable, the clay being described as laminated with silty sand laminae at 9.00–15.00 m in borehole 40 and similarly in borehole 42. In borehole 41, the grey clay is interbedded with sand units and no mention is made of laminations. However, this same borehole records organic 'peat' (detritus mud?) at 7.95–8.35 m. The overall thickness of these deposits reaches a maximum below the railway bridge adjacent to Pea Lane (TQ 583845), but they are cut out to the north and thin towards the south. They are absent south of borehole 32 east of Kemps (TQ 580834).

The undoubted lateral continuation of these deposits was seen in exposures made for a small lake in the Manor Farm field on the western side of Pea Lane. Although no exposures were available here during the present study, the farmer, Mr A. P. Mee, confirmed that the organic material had been recovered during the excavations. His descriptions were confirmed in borehole records he kindly supplied to the author. Immediately to the west, trial holes for a gravel pit around Stubbers Youth Camp record the deposits only in the north-eastern corner (borehole 7: TQ 575848), the remaining holes proving only 5–6 m of sand and gravel (equivalent to the Corbets Tey Member: section 2.5).

An exactly similar situation was seen in the Baldwin's Farm Quarry in 1978, where excavations revealed organic sediments at the eastern end of the faces 30 m south of Dennises Cottages. Here a sequence of organic and associated sediments formed a continuous accumulation that 'wedged in' above gravel and sand 4.5 m thick (section 2.5). The organic sediments were exposed for 26 m in a ENE–WSW direction and 70 m N–S. They comprised a basal green grey sand, 30 cm thick, resting on the gravels beneath. The sand was overlain by 85 cm of stiff light grey clay, with calcareous concretions. Above this, 70 cm of flat-bedded stratified sand occurred, with a laterally impersistent silt band 20 cm thick in the basal, laminated part of the sand. The section was capped here, as elsewhere, by brown pebbly clay 1.4 m thick, showing evidence of weathering and cyoturbation in the upper part. These organic sediments were sampled for palaeontological analysis (Chapter 6).

Trial boreholes for further excavations at Baldwin's Farm (kindly supplied by the manager, Mr Hudson) indicate that the organic deposits are only encountered in boreholes adjacent to the M25 motorway and Dennis Road, e.g. boreholes 5 (TQ 580836), 59–60 and 68 (TQ 577827). Similar holes for gravel extraction in a field east of the motorway, south of Kemps (TQ 584832), failed to confirm the deposits, but the records of these holes lack any details.

It would appear therefore that the organic sediments fill a channel-like elongate depression trending NNW–SSE beneath Manor Farm and Pea Lane and turning southwards beneath Kemps. It seems highly likely that this channel continues south to south-southwestwards and links with the deposits at Belhus Park (section 2.7g). It is conceivable that this channel-like infill may continue further to the south-west to link with the deposits at Purfleet, also described above. This conclusion is based on the stratigraphy and the altitudinal position of the deposits. Alternatively, it is possible that the deeper channel in the Pea Lane–Manor Farm area extends to the south-east, continuing north of South Ockendon, and rejoining the Mar Dyke valley course. Potentially this route would also make it highly possible that the Purfleet sediments are the downstream equivalent of the channel. The depression is here termed the Ockendon Channel.

Upstream the channel deposits may have been encountered during excavations reported by Ward (1984) at Park Corner Farm, Upminster (TQ 54978504: 15.3 m). Here excavation for a sewer revealed shell-rich material. A borehole at the site proved:

0–3.0 m	Brown clayey silt.
3.0–6.5	Light brown slightly clayey, sandy silt.
6.5–7.5	Medium coarse sand and gravel.
>7.5	London Clay.

Although there was no mention of shells in the borehole record, Ward concludes that the brown silt and sandy silt were the shell-bearing sediments. He also notes that the clayey silt was laminated in part. Ward also mentions that other boreholes along Park Farm Road prove the occurrence of 'a sequence of clayey and sandy sediments up to 10 m thick [occupying] a broad depression in the London Clay, extending down to 6 m OD about 350 m ESE of Park Corner Farm.'

It is not known how far the deposits extend to the north or west. However, Whitaker (1889) reports that 'in a well sunk at a depth of 30 ft (9.2 m) bones and shells were found in sand' at Cranham Hall (TQ 571861: 22.5 m). A further important exposure has also been discovered recently in Upminster (TQ 553861) and is currently under investigation.

2.7 AVELEY/WEST THURROCK/CRAYFORD SILTS AND SANDS

(i) Ilford

The Pleistocene deposits at Ilford have been famous for over 150 years, because of the remarkable number of finds of vertebrate remains and shells discovered in workings throughout the nineteenth century. There is an extensive literature on the area, examination of which reveals that there has been a complex of several sites, within a distance of over 2 km, from the bank of the Roding to Seven Kings (Fig. 1). These sites are summarised by Whitaker (1889), Hinton (1900) and Rolfe (1958).

The earliest record is of a large portion of an elephant skeleton found in 1824 by a local man, Mr Gibson, 'from the brick-field on the London turn-pike-road'; 'Mr Clift and Prof. Buckland being present at the exhumation' (Woodward & Davis, 1874, p. 391). Morris (1838) saw three sections, one 'adjoining the river Roding, the second about 500 yards from the river and the third beyond the town' on the north side of the High Street. The strata in each of these sites was almost the same. The first of these was almost certainly the famous Uphall Pit (TQ 436856: 9 m) at the northern end of Uphall Road, south of Ilford High Road. No description survives of the second, but the third seems to be the High Road site of Cotton (1847) and Hinton (1900).

The main descriptions of the Uphall sequence are given by Dawkins (1867) and Phillips (1871). A vast collection of vertebrates was made by Sir Antonio Brady from this site. Sections here seem to have remained open for several decades, but had been abandoned 'for some years' according to Whitaker (1889, p. 413). The sequence comprised:

0–0.9 m	Soil.
0.9–2.14	Irregularly stratified, contorted gravel and 'brickearth' with race. Pebble long axes orientated vertically.
2.14–2.82	Sandy loam.
2.84–4.06	Stratified and ripple-bedded yellow, ferruginous sand, with scattered gravel. This unit contained many shells and bones.
4.06–4.15	Irregular bed of gravel and sand. Associated with this bed was the famous *Mammuthus primigenius* skull.
4.15–5.98	Yellow, ripple-bedded fine sand interbedded with chocolate-coloured 'brickearth', containing layers of pebbles, clay bands and shell layers. The basal stratum of this unit contained abundant vertebrate remains, including the *Dicerorhinos hemitoechus* skull collected by Brady. (The so-called '*Corbicula* bed'.)
5.98–8.73	Gravel and sand.
>8.73	London Clay.

The sections are remarkably consistent in the various descriptions available.

The High Road section (TQ 447871: 13.5 m) is described in detail by Hinton (1900) and can be summarised as follows:

0–1.22 m	Clayey silty gravel ('trail'), with nearly all pebbles having long axes vertical.
1.22–2.44	Sandy loam, more argillaceous in places.
2.44–2.75	Inconstant yellow sand shell bed containing numerous shells and bones of *Bos primigenius*.

2.75–3.97	Stratified sandy clayey silt.
3.97–4.28	Shell bed, resting on an eroded surface of the underlying sediment.
4.28–5.80	Buff-coloured clayey silt and marl, with race; somewhat sandy (bone bed).
5.80–7.32	White sand with a few shells.
7.32–16.49	Coarse gravel and sand.
>16.49	London Clay.

The descriptions of this site by Cotton (1847) and Hinton (1900) record the variability of the sediments exposed in the different faces. In some places the individual beds changed facies. For example, Hinton (1900, p. 274) stated 'on cutting back the sand, it was seen that the 'bone bed' became more sandy and finally developed into an interstratified series of sand and marl.' He also described the overlying 'trail' in detail and interestingly noted that this material was much contorted, with 'furrows…8–9 inches (18–22 cm) in depth'. He also noted in the 'loam' and 'trail' 'suncracks…that consist of vertical fissures filled with sand.' He distinguished two distinct generations of these cracks: those that reach the surface he called 'recent' and those that do not, he called 'Pleistocene' (see below).

The stratigraphy of this spread of deposits seems to have been broadly consistent under the town. Hinton (1900, p. 274) remarked 'Dr Cotton (1847) noted a similar development of Drift occurring south-west of Uphall. This great development is influenced by the presence or absence of the underlying gravel is shown by the fact that between the two places the brickearth attains a thickness of about 20 ft (6.1 m) and that it rests directly on London Clay.' The last point was also mentioned by Woodward & Davies (1874, p. 393). Further detail of sections is given by Whitaker (1889, pp. 410–5). From these descriptions it also appears that contortion of the sediments to a depth of over 2 m is present everywhere. The sequence also apparently thins southwards, south of the High Road.

The development in the late nineteenth and early twentieth centuries prevented further observation of sections in the Ilford deposits for many years. However, the cutting of a sewer trench from Green Lane (TQ 447865) to Ilford High Road (TQ 446868), passing close to the former workings, provided new sections that were recorded in the late 1950s by Rolfe (1958). These sections from Gordon and Connaught Roads comprise (from south to north):

	Section I: 10.7 m OD
0–3.05 m	Yellow brown 'brickearth'.
3.05–3.96	White and yellow sands, occasionally well-stratified and containing many shells of *Corbicula fluminalis* and bones.
>3.96	Gravel and sand.
	Section II: 10.7 m OD
0–3.05 m	Yellow brown 'brickearth'.
3.05–3.75	Ferruginous sandy gravel.
3.75–3.84	*Corbicula* sands.
>3.84	Gravel and sand.

Section III: 11 m OD
0–0.91 m   'Brickearth' sandy at the base.
0.91–>3.35   Sandy gravel.

As Rolfe emphasises, these sections confirm earlier descriptions of the lateral discontinuity of individual beds.

This point is taken up by West *et al.* (1964) who describe a series of boreholes made for buildings immediately north of the railway line, next to Seven Kings Station (TQ 452871: 13 m). The detailed stratigraphy of the boreholes is given in the paper and is summarised here as follows:

0–0.61 m	Soil.
0.61–2.14	Brown red sandy clay and gravel.
2.14–3.05	Brown sand.
3.05–5.19	Brown red sandy silt and clay.
5.19–5.34	Brown red silty clay, muddy and shelly towards the base.
5.34–5.61	Brown grey clay mud with shells.
5.61–6.07	Brown detritus mud, with transition to bed above.
6.07–6.12	Grey clayey detritus mud with clay laminations.
6.12–6.75	Grey muddy silt becoming sandy in basal 20 cm, with transition to bed above.
6.75–9.09	Grey, becoming yellow stratified sand.
>9.09	London Clay.

The organic deposit clearly, therefore, occurs between a basal sand and the overlying clays ('brickearth') on the north side of the site, but it is missing to the south, the 'brickearth' lying directly on the sands. London Clay underlies the site at 3–4 m OD and is not overlain by the gravel and sand recorded commonly elsewhere.

The most recent observations in the area were made by the author in 1983–5. During investigation for this project, extensive excavations were made south of the High Road for relief road construction. These excavations ran east–west from Ilford Lane (TQ 436863) along Winston Way to a roundabout at Griggs Road (TQ 444865). Sections exposed at the junction with Richmond Road were sampled by A. Currant (British Museum, Natural History). The full sequence, as previously described by earlier authors, was exposed at the junction of Winston Way and Ilford Lane, but was unfortunately inaccessible. However, it comprised basal gravel and sand, overlain by *c*. 1.2 m of interbedded stratified sand and silt bands, 15–20 cm thick and was capped by 2.0 m of brown clay.

The 'brickearth' has been repeatedly encountered in shallow excavations at Ilford Cemetery (TQ 449867) where it was at least 1.7 m thick. At sites south of Ilford High Road in foundation excavations (TQ 439864 and 440864), particularly adjacent to Town Hall and Clements Roads, over 1.5 m of massive red-brown sandy clayey silt with irregularly distributed pebbles were seen. At the former site, the pebbles showed vertically orientated long axes (cf. previous descriptions) and included vertical wedge-like fissures filled by gravel and fine sand. The uppermost part of the fissures was overturned towards the south, indicating post-depositional movement. However, the most informative section was seen in the construction of a subway under Winston Way, at Riches (formerly Richmond) Road (TQ 442865). Here 2.5 m of sediment were visible and comprised massive brown silty clay. At about 1 m below the surface a 35 cm thick unit of crudely bedded, laterally variable silt and clay laminae 1–2 mm thick occurs. The upper 1.5 m of the decalcified, but carbonate, nodules are present below. However, the most striking features observed were narrow vertical dykes, regularly spaced every 1.2 m and infilled with vertically bedded sand. All the dykes seen reached the ground surface. These features, almost certainly sand wedges of periglacial origin, may be the 'suncracks' recorded from the High Road Pit sections by Hinton (1900; see above). Grain size distributions from this locality are given in Fig. 43 (see below).

The deposits underlying Ilford have also been recorded in boreholes for the relief road (Fig. 25). From these the disposition of the sedimentary units is clear. Immediately south of Ilford High Road, the shell-rich sands and silts are banked against a London Clay slope, possibly an old valley side inherited from the deposition of the underlying gravel and sand of the Mucking Member (section 2.6). The fossiliferous deposits reach a maximum thickness of about 4.5 m, are laterally quite extensive and are overlain throughout the area by the 'brickearth' already described. In the longitudinal section the relationship of the deposits to the Roding Valley is apparent. The deposits clearly decline towards the valley, where there is a precise correspondence in both height and sequence of borehole 34 and the Uphall Site. From earlier records it is clear that the latter occurred at a lower altitude than the sites adjacent to the High Road (Fig. 25). Indeed, the eastern valley side appears to represent a terrace underlain by these sediments. This latter observation is very important and was originally proposed by Woodward & Davies (1874, p. 392). This terrace has been confirmed by field observation (Fig. 1).

Taking the deposits as a whole, therefore, there are unquestionably three distinct units at Ilford: the gravel and sand of the Mucking Member, the fossiliferous Ilford Sands and Silts (new member term proposed here: stratotype section, borehole 30, Seven Kings) and the 'brickearth'. The latter should not be confused with the 'Ilford Brickearth' of some authors to which fossil finds have generally been wrongly attributed in the past. The Sands and Silts, together with the organic sediment at Seven Kings, appear to represent a discrete sedimentary unit comprising a variety of beds, but being of fluvial origin. The unit clearly grades from ENE to WSW, the sediments at Uphall representing an aggradation associated not with the Thames, but with the Roding. Indeed, the conclusion of West *et al.* (1964) that the deposits represent the Seven Kings Water stream is almost certainly correct. From this evidence the Ilford Sands and Silts unequivocally post-date and rest on a surface partially incised into the Mucking Gravel and the London Clay. In turn, they predate the 'brickearth' which is almost certainly the lithological equivalent of the Langley Silt Complex (section 2.10).

(j) Hackney Downs, Stoke Newington and Highbury

The earliest observations of Pleistocene deposits in the Stoke Newington area were by Prestwich (1855) who recorded a section in Shacklewell Lane (TQ 33958600: *c.* 21 m). Here large vertebrate bones were recovered from clay. The full sequence from the site was:

0–0.91 m	'Brickearth'.
0.91–2.74	Flint gravel and irregular layers of yellow sand.
2.74–3.50	Dark grey sandy clay with plant remains, tree trunks, some bones (at the base) and numerous shells.
>3.50	Light yellow sand and ferruginous gravel (base not seen).

The abundant shells, plant remains and the remains of an 'ox' (?*Bos* sp.) were all identified and Prestwich concluded that they represented quiet water sedimentation in a shallow stream under conditions similar to the present day.

Fourteen years later a similar sequence was reported by Tylor (1868) from Highbury Park. Here two brickpit sections revealed important thicker sequences. The more westerly of these was probably immediately south of Kelross Road at TQ 322858 (37 m), whilst the eastern pit was between Collins and Catherall Roads (TQ 324860: 32 m). The former seems to have exposed 12.25 m of 'brickearth', rested on 3 m of unfossiliferous current bedded yellow sand with purple clay bands, containing wood fragments, beneath. At the latter, a clay bed 62 cm thick was found, containing abundant land and freshwater shells, together with wood. The clay was immediately overlain by reddish 'loam' or 'brickearth'. The London Clay was present 3 m below the clay which presumably rested on gravel and sand. These deposits were correlated by Tylor with those at Shacklewell, both on the basis of their fossil contents and their stratigraphical position. In addition, he also noted some earlier finds of shells in sediments exposed in a cutting (?for a railway line) on Hackney Downs in 1866 (see below). Here sediments with *Corbicula* and *Unio* had been exposed at about 14.5 m OD.

Of particular significance in the Stoke Newington–Hackney area were the discoveries of the amateur archaeologist Worthington Smith. The rapid expansion of Victorian London in the second half of the nineteenth century resulted in large-scale house building in this area. Smith, a civil engineer, watched the excavations for the basements of houses and sewerage in the area and noticed that artefacts in sharp condition were present in and immediately beneath the 'brickearth' of the district. This led him to propose that a 'Palaeolithic floor' existed in the 'brickearth' sediments of north-east London. As work progressed, Smith and Greenhill collected hundreds of worked pieces between them. Their finds were reported in several publications (e.g. Smith, 1879, 1884, 1887, 1894; Greenhill, 1884). For more recent summaries see Wymer (1968), Roe (1981) and Harding & Gibbard (1984).

Fortunately, both Smith and Greenhill were careful to describe the geological situation of their finds. It appears that the main 'floor' occurred in the 'brickearth' and not under it. This horizon was actually a thin bed of subangular gravel or a colour contrast. However, in some areas a second, lower 'floor' was recognised on the surface of a coarse gravel bed, particularly in the area immediately north of Stoke Newington Common. This bed often also seems to have included the remains of large vertebrates. Between the gravel and the 'brickearth', Smith clearly illustrates the occurrence of a unit of bedded yellow sand over 2 m thick. The upper part of this sand appears to have been greatly disturbed by cryoturbation, the 'brickearth' sediments having been injected into the sand beneath in drip and wedge-like (?thermal contraction crack) structures. Some of these structures illustrated by Smith, closely resemble low-angle flow contortions, resulting from downslope movement of the overlying 'brickearth' sediment. According to Roe (1981, p. 175), Warren relocated the 'floor' at Geldestone Road (TQ 343870), but his finds were not published.

In addition to these sediments, Smith also recorded shell-bearing sediments, including species recorded at the sites already mentioned, at or close to Clapton Railway Station (c.TQ 347865). This section is illustrated in Smith (1894, fig. 139). Here shelly yellow sand 30 cm thick rested on gravel and was overlain by up to 1.5 m of 'brickearth'. Local sections were also recorded by Whitaker (1889, pp.402–4). West of the High Street, between Church Street and Victoria Road (c.TQ 334862), he observed 0.60–1.25 m of shell–rich stratified brown clayey silt, overlain by 2.1 m of cross-bedded sand and gravel and capped by over 2.5 m of brown 'brickearth'.

The completion of development in the area and the exhaustion of the 'brickearth' spreads led to a cessation in detailed investigations of the area for almost a century. Although much discussed (e.g. by King & Oakley, 1936), the deposits in the area were not studied until excavations were undertaken in 1971 by Sampson and Campbell on Stoke Newington Common (according to Roe, 1981). However, the results were never published. Roe also reports that test pits were dug on the north side of Cazenove Road in 1975, but 'no sign of the floor was revealed'. However, in 1981 the demolition of a house at 55 Northwold Road (TQ 33988663: 22 m) provided an opportunity to undertake detailed investigations both at the site itself and in the vicinity. The results of the site excavation were presented by Harding & Gibbard (1984), but the regional study has not been previously published.

At the Northwold Road site, Harding & Gibbard (1984) found a 1.1 m thick unit of gravel and sand resting on a dissected London Clay surface. This was overlain by 1.5 m of massive clayey silt, in turn overlain by 0.8 m of yellow-brown silt. Detailed analyses indicated that the gravel and sand was of fluvial origin, whilst the clayey silt above was a colluvial (solifluction) deposit. The upper silt was fine-grained alluvium. Although two Palaeolithic artefacts were found during these investigations, it was concluded that the 'floor' had not been encountered.

The accompanying regional study, however, greatly clarified the geological sequence in the area. This was undertaken using borehole data from various sources and holes drilled by the author.

Previous reports indicated that the main sequences were aligned along the slope from Stamford Hill to the lower ground on which Stoke Newington Common stands. A series of north–south aligned transects was therefore undertaken perpendicular to the slope (Fig. 18). The furthest west extended from 43 Cazenove Road, through the George Downing Estate to the Common. The second was located in the gardens of flats in Kyverdale Road and beneath house 14. The third series of holes was put down for a garage behind 2 Fountayne Road. The next was located in the grounds of Inglethorp House (Tower Gardens Estate) in Geldeston Road and the last was put down in the grounds of the Evering Road flats.

These boreholes proved a consistent sequence comprising a yellow to grey sand unit overlying gravel and sand and capped by mottled grey-brown to orange very clayey silt to clay with small stones ('brickearth'). Since almost all the records of these sediments are from boreholes, it is not possible to describe the sedimentary structures or the stratification in detail. However, from the recovered sediment, it is obvious that this sand unit (termed here the Highbury Silts and Sands Member) is internally complex. It includes a number of grey to green or brown silt bands, each generally only a few centimetres in thickness, at various levels. Some thicker bands also occur: for example, in borehole GD1 a 20 cm thick grey clay band was present. The sediments are clearly finely stratified in part, occasionally showing internal erosional conntacts on lower subunits (possibly small channel-like scours) and include occasional, thin, fine pebble bands. Nowhere could the character of the basal gravel be determined during this study; however, earlier boreholes have proven the gravel to be a maximum of 3.5 m thick, although usually much less, and it rests directly on London Clay. No fossil remains were recovered from any of these sediments.

The sands and underlying gravel abut the Stamford Hill Gravel along a WNW–ESE trending line that closely parallels the 25 m contour from Cazenove Road to Rossington Street.

This sequence closely matches the descriptions and illustrations of the sand and overlying 'brickearth' by Smith (1894 etc.). As mentioned above, he clearly shows that the sand was interbedded horizontally with finer sediments. The modern results fully support his observations. The sands and the underlying gravel seem to represent the infilling of a channel trending broadly west–east beneath the sloping ground between 26.5 m and 19 m OD. The continuity of this unit further west is proven by boreholes put down in Abney Park Cemetery, close to Stoke Newington Church Street (TQ 333866). Towards the east, it appears that the shelly sands near Clapton Station (c. TQ 347865), described by Smith (1894), may be the lateral equivalent of the Highbury Silts and Sands Member. On the basis of the cross-sections, the Highbury Member is cut out by the modern ground slope to the area of Stoke Newington Common, associated with the valley of the Hackney Brook stream. This stream used to flow south-eastwards across the High Street at its junction with Northwold Road. It then turned east of the High Street, then south-east again south of Hackey Downs (see below).

The area south of Stoke Newington Common to Hackney Downs has already been partly discussed above (section 2.5 and 2.6). Transects across the area (Figs. 16, 18, 27) indicate that an extensive spread of gravels occurs south of Stoke Newington Common to immediately north of Hackney Downs Station. This spread also extends from the Lea Valley, where it is cut out by later erosion by the River Lea and to Highbury, where it abuts the London Clay hill (Fig. 10). As already mentioned, the gravels are overlain by organic sediments and these in turn are succeeded by further (later) gravel, the latter presumably resting directly on the lower gravels where the organic sediments have been cut out or are absent. That the unit was deposited by a southward flowing stream is reinforced by contours drawn on the basal surface of the organic sediment surface in the dense network of boreholes at Nightingale Estate (Fig. 28). As has already been shown, this lower gravel and sand unit is almost certainly the Lea Leytonstone Gravel. The upper gravel and sand unit is here termed the Hackney Downs Gravel (stratotype TQ 345861: 21 m).

Organic sediments were encountered beneath Hackney Downs in boreholes for the Nightingale Estate flat blocks. Here over 70 boreholes record up to 2 m of dark grey silty clay with plant remains, wood and shells. In order to sample this sediment two boreholes were sunk immediately north of the service road (Muir Road), south of Kenninghall Road (NE1: TQ 34578613 and NE2: 34518612: 21 m). The records were as follows:

NE1

0–1.60 m	Made ground.
1.60–1.90	Green-grey pebbly silty clay.
1.90–4.10	Orange-brown sand and gravel.
4.10–4.65	Dark grey silty clay with occasional pebbles and shells.
4.65–6.10	Brown clayey gravel and sand.
>6.10	London Clay.

NE2

0–0.80 m	Made ground.
0.80–4.30	Orange sandy gravel.
4.30–4.98	Dark grey silty clay with plant remains, shells and isolated stones.
4.98–5.70	Grey to light grey silty clay with shells and occasional pebbles and sand laminae.
5.70–6.10	Brown gravel and sand.
>6.10	London Clay.

These boreholes were sampled for pebble counts and palaeontological analyses (Chapter 6). There can be little doubt that these organic sediments are the lateral equivalent of those seen by Tylor (1868) in the adjacent railway cutting. The base of the overlying gravel and sand unit (Hackney Downs Gravel) appears to rest on an eroded surface on the organic sediment, the latter being cut out in places completely. On the basis of contours on the base of the organic sediments beneath Nightingale Estate, the sediments appear to gently decline in altitude from the north-east towards the south-west. Tracing the sequence eastwards shows that the deposits described by Prestwich (1855) from Shacklewell (see above) and indeed those further west beneath Highbury are undoubtedly the same unit (Fig. 18), as proposed by Tylor (1868).

Three boreholes put down for the building of St Augustine's Vicarage, Highbury New Park (TQ 340857: 31 m) penetrated these deposits adjacent to Tylor's (1868) original sites. These holes record:

0–8.40 m	Made ground (? 'brickearth' removed).
8.40–10.10	Yellow to brown banded silty clay with some shell fragments.
10.10–11.30	Black amorphous 'peat' (clay mud).
11.30–13.30	Green-grey calcareous silty clay becoming grey and more shell rich with depth.
13.30–16.50	Yellow-brown sand and fine gravel.
>16.50	London Clay.

Samples for palaeontological analysis, taken from 10.10–13.30 m (borehole 1) and 11.00–13.00 m (borehole 2), were kindly supplied by J. A. Zalasiewicz (British Geological Survey) (Chapter 6). As already noted, these sediments appear to abut the eastern side of Highbury London Clay hill, and in Fig. 18

they apparently occupy a large meander-like bend excavated into the bedrock. The boreholes in this section record yellow sand similar in all respects, including height, to the Highbury Member. On this basis there appears to be no reason not to correlate these deposits. If this correlation is correct, it clearly demonstrates that the Hackney Downs Gravel must post-date accumulation of the Highbury Sand and its basal gravel. However, it is possible that the basal gravel may predate the Hackney Downs–Arsenal organic deposits, in the absence of organic sediments north of Stoke Newington Common. Correlation of the sand at Stoke Newington and Highbury implies that the organic sediments throughout the Hackney Downs–Stoke Newington area were originally buried by 2–3 m of sand that has subsequently been removed. This removal must therefore have taken place either before or during the emplacement of the overlying Hackney Downs Gravel.

From the earlier reports and modern borehole records, it is apparent that the 'brickearth' sediment originally blanketed virtually all the deposits in this area. Exploitation of this sediment has given a somewhat unrepresentative aspect to the modern sequence. Nevertheless, enough material remains for an appreciation of its deposition and possible origins (Chapters 6, 8). In the Stoke Newington area, particularly north of the Common, it is clear that the sediment is largely of solifluction origin, the material having been presumably derived from upslope. Therefore the evidence collected during this study indicates that Smith's 'floor' is in reality material reworked from Stamford Hill. This interpretation explains the known distribution of the finds and explains their apparent condition. Indeed, the gravels beneath Stamford Hill include artefactual material (Wymer, 1968). The so-called 'lower floor' in the basal gravels of the Highbury Sand unit presumably represents earlier reworked material incorporated into the gravel as a 'lag' type assemblage. It is, of course, possible that the material may have been present on the ground in the area before accumulation of these basal gravels.

As already mentioned above, the sequence of sediments described has been incised by the Lea tributary, the Hackney Brook. Later floodplain sediments that occur in the small valley cut by this stream were also described by Smith (1894) immediately east of Stoke Newington High Street (see Wymer, 1968, pp. 297–302, for summary). The most important site is that immediately west of Bayston Road (TQ 337862) from which two birch stakes were found. This site was not available for study during the present work. However, from the disposition of the deposits they must post-date much of the deposits described here. Moreover, it appears that the sequence may be equivalent to that present at the Northwold Road site and beneath the Common. The transverse borehole sections demonstrate the relationship of the Hackney Brook valley to the older sediments (Figs.16, 18, 20).

In summary, since the lower gravel and sand unit beneath Hackney Downs has been equated with the River Lea Leytonstone Gravel, it seems highly likely, in view of the general southward decline and disposition of the overlying Hackney Downs Organic Deposits, that the Highbury Sand and the Hackney Downs Gravel were all laid down by the River Lea rather than the Thames. Subsequent abandonment of the area by the Lea allowed the accumulation of 'brickearth' sediment and development of the Hackney Brook tributary.

(k) Camden Town

Whitaker (1889, p. 401) reports a find of 'bones of *Elephant* and *Hippopotamus*' collected from 'Copenhagen Fields' by a Mr Wetherell of Highgate. According to a note of 1857: "Fragments of bones...were found...in digging for a sewer on the side of the road leading from Holloway to Camden Town, and near Brecknock Crescent. The bed in which they occurred was only a few feet from the surface, and rested on London Clay.' From this description it seems that the localities were on the north side of Camden Road, near the junction with Brecknock Road (TQ 298850), i.e. close to the western limit of the 'brickearth' spread associated with Islington and Highbury Hill (discussed in the preceding section). Presumably the bone-bearing stratum underlay the 'brickearth'. Nothing further is known about this site or the fossil remains; however, if these fossiliferous sediments were of fluvial origin, they may have been associated with the eastern branch of the River Fleet. Re-examination of these sediments would be worth while.

(l) Peckham

During resurveying of the South London geology map, F. Berry (BGS) discovered organic sediments in boreholes in the Peckham area. Boreholes put down at Harder's Road (TQ 34647649: 7.65 m) in the early 1970s to investigate the sediments yielded the following sequence (details from BGS borehole record, supplemented by information kindly supplied by R. Beck):

0–1.07 m	Made ground.
1.07–1.47	Sandy clay.
1.47–2.31	'Cryoturbated' gravel.
2.31–5.31	Stiff clay with shells and organic detritus in parts, becoming more organic towards the base.
5.31–5.64	'Peat'.
5.64–5.99	'Peat' and detritus mud.
5.99–6.71	Detritus mud with wood and rhizomes.
6.71–12.30	Olive-grey laminated clayey silt to silty clay with shells.
12.30–12.34	Fine gravel. (Not bottomed.)

The regional setting of the Peckham sequence is shown in Figs. 34 and 35. Here the deposits appear to fill a depression cut into Reading Beds to a maximum depth of −7.75 m OD. The basal sediments are gravel and sand 2.75 m thick. This unit appears to thicken northwards to reach a maximum of 12 m in borehole 673 south of Peckham High Street. To the west the gravel and sand is variable in thickness and can be traced at least as far as St Giles Hospital, Camberwell, where it reaches 3.6 m.

Overlying the gravel and sand is an internally variable unit that is the equivalent of the clays and organic sediments in the BGS boreholes. In spite of their variation in thickness they can be traced beneath Peckham Road/High Street to Camberwell. Internally, the unit is similar to that described from Harder's Road. It comprises a lower shell-bearing, laminated clay, which is underlain in places by sand 1.2 m thick. The clay is restricted to the immediate area of Harder's Road, but is also recorded beneath St Giles Hospital, where it is 3.8 m thick. In both these areas organic sediment or 'peat' (?detritus mud)

2.7 AVELEY/WEST THURROCK/CRAYFORD SILTS AND SANDS

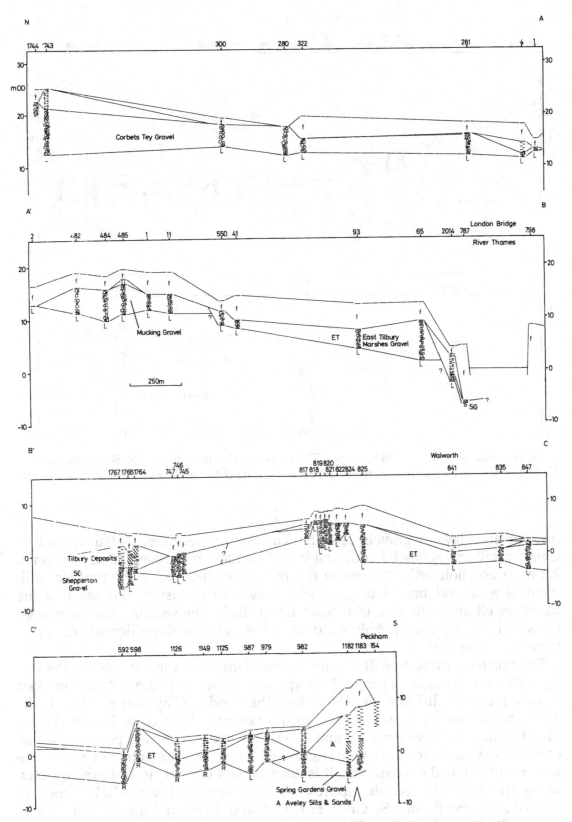

Fig. 34. Section from London Bridge to Peckham, constructed using borehole data (sources: GLC, BGS).

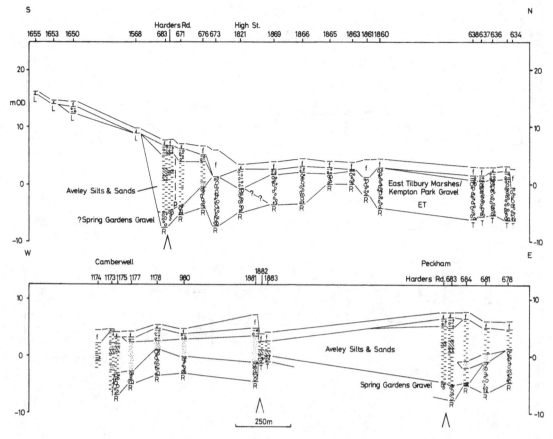

Fig. 35. Sections from the Peckham area. The upper section is orientated N–S and the lower W–E. The sections were constructed using borehole data obtained from BGS and GLC sources.

rests on the clays or gravels beneath. The organic sediment is 0.6 m thick at Camberwell and beneath Lyndhurst Way, Peckham, but reaches a maximum of 2.2 m in borehole 683, adjacent to the BGS borehole. Above the organic sediment is a second brown to grey silty clay, which passes into sand in the Camberwell area, where it is at least 1.2 m thick. The sections are capped by brown clayey sand or pebbly sand or pebbly clay, a slope deposit common throughout the area.

The north–south section illustrates the relationship of the sequence to the valley side and younger deposits. It is apparent that the sediments are banked against a buried cliff-like feature cut into the London Clay and Reading Beds. To the north and north-west they abut the Kempton Park Gravel/East Tilbury Marshes members (sections 2.9 and 2.11). However, the actual position of this contact is difficult to determine accurately from the borehole records in the absence of detailed descriptions. It is nevertheless possible to trace the contact, using the borehole records, approximately along Peckham High Street, to immediately north of St Giles Hospital and then to Camberwell Grove, Denmark Hill (TQ 765329). The deposits are also probably cut out south of Queen's Road, Peckham railway station. The sequence can therefore be traced a distance of c. 2 km east–west and 400 m north–south.

2.8 West Thurrock Gravel

As already mentioned (section 2.7c), sections in the Lion Tramway Cutting, West Thurrock, immediately north of the London Road bridge (stratotype: TQ 59777792: 8 m) expose horizontally bedded medium gravel and sand. This deposit clearly abuts, cuts out and therefore post-dates the sand and 'brickearth' sequence exposed further north in the cutting sections (cf. Hollin, 1977). A pebble count from this unit is given in Appendix 2. Boreholes in the immediate vicinity of the Tramway Cutting (Fig. 31) confirm the relationship of the units described. They indicate that the gravels and sands are 9 m thick and rest on bedrock at –4 m OD. The deposits are truncated to the south by downcutting to a level below the Tilbury Deposits (section 2.13). To the west the spread continues to the neighbouring cutting made for access to the chalk quarry east of Sandy Lane, West Thurrock (section 2.7c). Here the exposures showed an identical relationship to that in the Lion Cutting.

Towards the east the gravels appear to thin. However, in the Little Thurrock (Grays Thurrock) workings, the 'brickearth' and associated deposits are truncated on the south side by coarse gravel and sand that appear to represent the lateral continuation of the West Thurrock Member (section 2.7b). The full thickness of these deposits was seen by Tylor (1869, p.84) who described them as 'a section of coarse gravel 12 ft. (3.66 m) thick, open in a gravel-pit, reposing on chalk without any beds of sand or brickearth. The chalk is…9 ft. (2.75 m) above the rails at the bridge' (Fig. 31).

On the basis of altitude and stratigraphical position, it is possible that the gravel, present immediately adjacent to the Erith–Crayford 'brickearth' deposits beneath Slade Green, is the lateral equivalent of this unit. According to Chandler (1914) '10 to 15 ft (3–4.5 m) of sandy gravel was exposed' at the Howbury Pit (c.TQ 526767: 5 m). No exposures were available during the present study. If this correlation is correct, the stratigraphical relationship in the Crayford area is identical to that at Grays and West Thurrock. Similarly, deposits of this member are probably present at Aveley, north of Sandy Lane in boreholes for the A13 (Fig. 7) at the appropriate height, from Rainham (Common Sewer) to near Wennington Hall, where they abut the Aveley Silts and Sands deposits (section 2.7d). The deposits are also present beneath the neighbouring field, a partially infilled quarry, at TQ 543817 (8.5 m) where they are 6 m thick, but thin rapidly eastwards.

Because of the fragmentary nature of this unit and its close similarity in both altitudinal distribution and lithological composition to that of the preceding Mucking Member, it is impossible to separate these two spreads in the absence of intervening deposits. This description therefore represents only those localities where this unit can be reliably identified. Potentially the West Thurrock Member may be much more extensive in the Thames Valley. However, a long profile of known deposits is shown in Fig. 37 (see below).

2.9 Kempton Park/East Tilbury Marshes Gravel

As previously noted by Gibbard (1985, p. 62) the term 'Floodplain gravel' was apparently first used formally by Bromehead (1912, p. 75) to include 'all river gravels younger than the Taplow' (Dewey & Bromehead, 1915, p. 76). The famous twofold division into Upper and Lower Floodplain gravels separated by an erosional episode was proposed by Dewey & Bromehead (1921). The recent work of Gibbard (1985) confirmed that the unit underlying the 'Upper Floodplain Terrace', the Kempton Park Gravel Member, could be recognised from near Marlow, Buckinghamshire, to central London. In the Trafalgar Square–Whitehall area deposits of this unit abut Ipswichian Stage freshwater and estuarine sediments and underlie Horse Guards Parade. Similarly they occur south of the river in the Clapham area.

Radiocarbon dating of organic material from several localities in this Member indicates that it accumulated between about 45 000 and 30 000 BP, i.e. the Middle Devensian Substage. The gravels and sands of this unit contain numerous vertebrate remains in places, but are poor in archaeological artefacts.

Bridgland (1983a, 1988a) recognised a unit stratigraphically younger than his Taplow-equivalent Mucking Gravel east of Tilbury that he termed the East Tilbury Marshes Gravel, based on a type section at TQ 688784 (0–1 m OD). He correlated this unit with the Middle Thames Kempton Park Gravel on the basis of downstream extension of unit gradient. This correlation has subsequently been confirmed by Gibbard *et al.* (1988).

In the East Tilbury Marshes area, gravel quarrying (TQ 688784) in the Thames' floodplain has shown a unit of gravel and sands disposed into braided river type facies. The type exposure comprises a large working in which occur up to 5 m of predominantly horizontal or very gently dipping beds of medium to fine sand matrix supported gravel 20–30 cm thick. Occasional coarser clasts are also present in places. The sediments rest on an undulating surface of Thanet Sand. Laterally impersistent sand lenses up to 50 cm thick, although more normally only a few centimetres thick, are also present. At one point an intraformational ice wedge cast 1.25 m long and 15 cm wide cut through the sediments. Palaeocurrent measurements from the sediments indicated flow towards 25° and a comparable gravel imbrication direction towards 20°. Two gravel samples were collected from this section (Appendix 2). The gravel and sand sediments here are overlain abruptly by 80–90 cm of massive grey silty clay. These sediments are almost certainly modern floodplain deposits. Dewey *et al.* (1924) previously reported gravel excavation on both sides of the East Tilbury Road here.

Because this unit underlies the modern floodplain, it is seldom exposed in the eastern part of the valley. However, it can be traced as remnants upstream. For example, it is present as a minor remnant at Grays adjacent to the river floodplain between −5 to 1 m OD (Fig. 36). At Stone Marshes (TQ 567753: 4 m) a section exposes 4 m of deposits. At the base, 2 m of medium stratified gravel with a sand and fine gravel matrix occurs and is overlain by 1.5 m of silty sand. This deposit includes a number of vertically orientated clasts and passes upward into a brown pebbly clay 50 cm thick. The same spread is shown by the borehole section along the Dartford Tunnel southern approach road (M25: Fig. 6), in which up to 10 m of gravel and sand underlie both floodplain alluvium and brown sandy silty clay with stones (slope diamicton or 'head'). This section clearly demonstrates the relationship of the member to the Shepperton Gravel and the valley side in this area.

Few boreholes prove deposits at the corresponding height for correlation with this unit upstream as far as the City. One example may occur at Erith

2.9 KEMPTON PARK/EAST TILBURY MARSHES GRAVEL

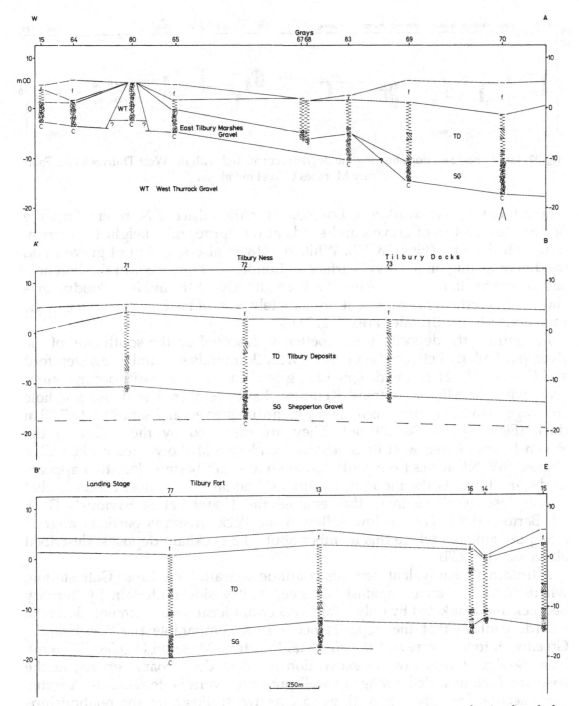

Fig. 36. Section along the River Thames bank at Tilbury Riverside, constructed using borehole data (source: BGS).

(TQ 507787: 3 m), where a borehole penetrated 1.12 m of gravel and sand resting on Thanet Sand. On the basis of the long profile (Fig. 37), the deposits occur at only 1–1.5 m OD in this area and would therefore be very difficult to separate from the modern floodplain. It may be represented on the north side of the river as a dissected spread immediately adjacent to the modern floodplain from south Hornchurch to Newham; it is probably present at Barking, Castle Green (TQ 468838: 4.5 m). The deposits recur beneath Newham (TQ 414822: 5 m)

2 LITHOSTRATIGRAPHY

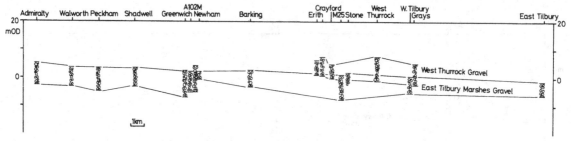

Fig. 37. Long profile section showing the deposits correlated with the West Thurrock and East Tilbury Marshes Gravel members.

where there are old workings. Boreholes for the adjacent Northern Outflow Sewer confirm 4 m of gravels and sands at the appropriate height for correlation with this unit (Figs. 38, 39). Whitaker (1889) also saw 3 m of gravels and sand here resting on a 'wavy' surface of London Clay in 1863. In the borehole section the relationship of the unit to those underlying the modern floodplain is shown. It is certainly also present immediately west of the Lea confluence in the Limehouse–South Bromley area (Fig. 21).

In contrast, the deposits are far better represented on the south side of the river, particularly between Southwark, west Bermondsey, Camberley, Deptford and Greenwich.. Here a wide spread of gravel and sand deposits occur, continuous with the Middle Thames' Kempton Park Member. The dense borehole coverage in the area confirms that they comprise gravels and sands up to 7.25 m thick (Figs. 34, 35, 38, 39, 40). They are dissected by the valley of the Ravensbourne River, west of Greenwich, and by a shallow stream-like valley aligned SW–NE across from south Lambeth to south Bermondsey that appears to decline towards the modern floodplain. The latter is almost certainly that of the 'lost river' Nekinger that entered the Thames at St Saviour's Dock (cf. Barton, 1962). The shallow valley of the Peck stream is partially aligned along the junction where this member abuts the Peckham deposits, described above (section 2.71).

Altitudinally equivalent deposits continue beneath New Cross Gate station, where they are banked against the steep valley side underlain by Tertiary bedrock and blanketed by only a thin veneer of Pleistocene sediment. Borehole records confirm that the deposits occur in a comparable position beneath Greenwich town centre and the National Maritime Museum (TQ 386777: 6 m). Here Boulger (1876) saw an excavation in 10 m deep coarse gravel, at the base of which included a range of well-preserved vertebrate remains. A transverse section from the Isle of Dogs to Greenwich illustrates the relationships in this area (Fig. 41). A maximum thickness of 15.5 m is recorded at the northeast corner of Greenwich Park (TQ 391778: 12m). The deposits continue to beneath the A102(M) road (Fig. 42), but are cut out by the floodplain 500 m to the east.

The longitudinal gradient obtained by plotting all the localities mentioned is 0.3–0.4 m km^{-1}, precisely that obtained for the Kempton Park Gravel upstream (Fig. 37. Unfortunately, the poor exposure of these deposits prevented pebble count samples being collected, except at the type section (Appendix 2). They comprise 91–93% flint, 2–3.5% quartz, 1–2% quartzite and 2–3.5% Greensand chert.

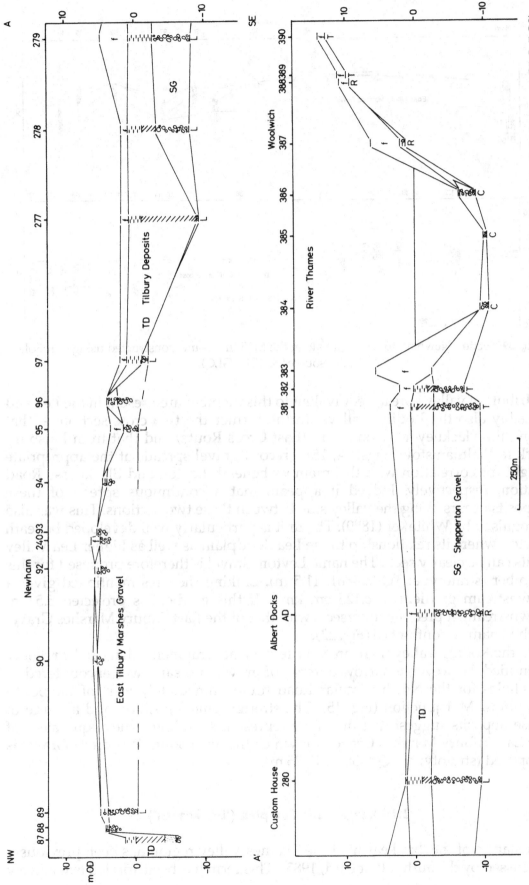

Fig. 38. Section along the line of the Northern Outflow Sewer from Newham to Woolwich, constructed using borehole data (sources: GLC, BGS).

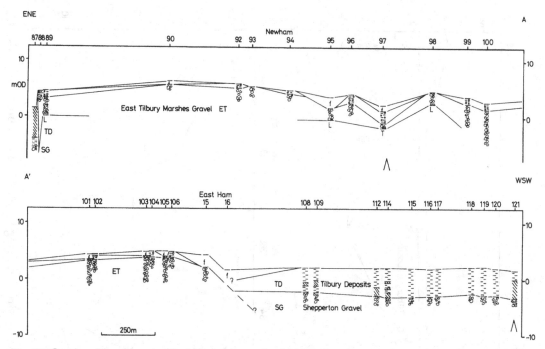

Fig. 39. Section Newham to East Ham along the M15 link route, constructed using borehole data (sources: ERCU, GLC).

Tributary valley spreads equivalent to this member are present in the Lea and possibly also the Roding valleys. In the former the two cross-sections – that along the Hackney–M11 link road (East Cross Route) and that from Finsbury Park to Walthamstow (Figs. 14, 15) – record gravel spreads at the appropriate height for correlation with this member beneath Leyton and Blackhorse Road Station, respectively. Indeed it appears that a continuous spread of these deposits occurs along the valley side between these two sections. This was also recognised by Whitaker (1889). The unit is particularly well developed beneath Leyton where its relationship to the Lea floodplain as well as higher Lea Valley units can be clearly seen. The name Leyton Gravel is therefore proposed for this member (stratotype: TQ 384861: 11.5 m). Linking the sites mentioned gives a downstream gradient of $c.125$ cm km^{-1}. If this gradient is projected 5.5 km downstream, it precisely intersects with that of the East Tilbury Marshes Gravel at the Thames' confluence (Fig. 22).

In the Roding Valley there are very few terrace fragments that can be reliably identified. However, a narrow outcrop of gravel and sand was encountered in boreholes for the North Circular Road route immediately west of the South Woodford M11 junction (Fig. 15). The stratigraphical position and altitude of these deposits suggest that they may represent a Roding Valley equivalent of the East Tilbury Marshes Gravel, for which the term South Woodford Gravel is proposed (stratotype: TQ 413884: 15.75 m).

2.10 Langley Silt Complex ('brickearth')

The nature of the 'brickearth' of the Thames Valley region has been previously discussed by the author (Gibbard, 1985). This term has been applied extensively

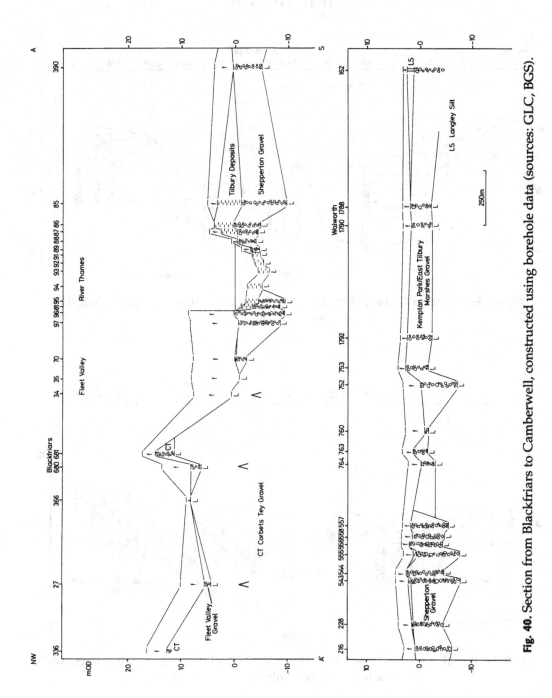

Fig. 40. Section from Blackfriars to Camberwell, constructed using borehole data (sources: GLC, BGS).

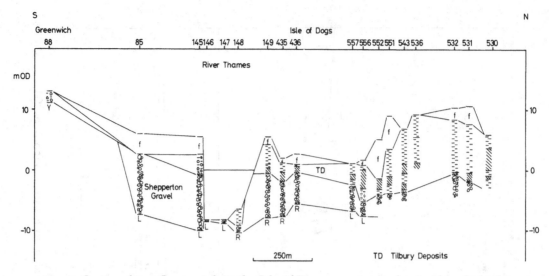

Fig. 41. Section from Greenwich to the Isle of Dogs, constructed using borehole data (source: BGS).

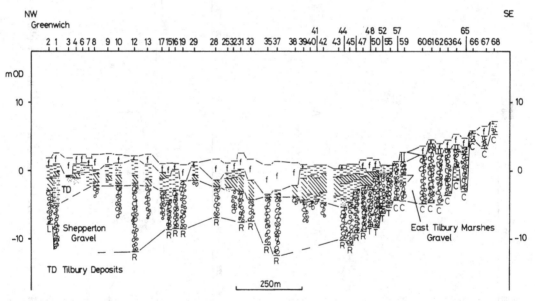

Fig. 42. Section along the A102(M) road, Blackwall Tunnel south approach at Greenwich, constructed using borehole data (source: GLC).

in the region to any fine-grained deposit that was suitable for brick-making and as a result it is very variable in character (cf. Bromehead, 1925). Throughout the area it varies from a sandy silt to remobilised Tertiary clay. There has been much debate over its origin and it is now generally accepted that the most abundant type, clayey silts, were formed as a combined loess and waterlain or colluvial deposit. However, it is important to note that this material differs in origin from the laminated or massive fossiliferous 'brickearths' of the area (cf. Hollin, 1977; section 4.2 below).

In the Middle Thames area Gibbard (1985) concluded that the silt component may be most commonly of loessic origin. This was based on their massive appearance, the vertical columnar desiccation cracks, the abundance of the sediments' characteristic tubule-like internal structure, decalcification and the presence of calcareous nodules (*loess pupchen*). However, the bimodality of grain size samples implies that the sediments are not pure loess but have been modified by slope wash. The latter is particularly common at the bluff or backslope of a gravel aggradation, where lamination in the basal part of the complex demonstrates the role of water during deposition.

Apart from loessic material, the name 'brickearth' has also been applied to demonstrably solifluction deposits, stratified esuarine or alluvial deposits (see above, section 2.7) and possibly *in situ* sediment the upper surface of which has undergone clay enrichment during palaeosol development. The latter is clearly not a sedimentary unit; however, the other types are.

Because of its polygenetic origin it is found in apparently conflicting contexts. The main mass overlies the Taplow and Lynch Hill-equivalent Corbets Tey and Mucking Gravels respectively. However, a veneer is also present on the Kempton Park/East Tilbury Marshes Gravel and the Aveley/West Thurrock/ Crayford sediments. It was earlier thought by some writers to be 'intimately associated' (Sherlock & Noble, 1922) or 'nearly contemporaneous' (Dines & Edmunds, 1925) with the gravel on which it rested by analogy with the modern floodplain alluvium. However, since the 'brickearth' can be generally shown to be a cold stage deposit, it is certainly not directly comparable to modern alluvium.

The term 'brickearth' cannot therefore be used as a precise descriptive, lithological or genetic term. For this reason the term 'silt' will be used here, as in the Middle Thames, to describe the dominant lithology. The single name 'Langley Silt Complex' was proposed for this purpose by Gibbard (1985, pp.56–7) and applied to the main thickness of deposits that were thought to be in their original position. The large working quarry at Langley (TQ 002800: 28 m) is the type section of this member.

As before, the approach adopted here for the correlation of this unit is largely descriptive, supported by grain-size analyses. The results of analyses of samples taken for heavy mineral and thermoluminescence (TL) dating are not available at the time of writing. However, in the Middle Thames a consistent series of TL dates were obtained by Gibbard, Wintle & Catt (1986) grouped around *c.* 17 000 y BP from the main mass of silt-rich sediment. More variable ages are found from the colluvial sediments (e.g. over 140 000 y BP at Yiewsley: A. G. Wintle, personal communication). Wintle (1982) and Parks & Rendel (1992) obtained a TL date on silt-rich sediment beneath the interglacial sediments at Northfleet of 150 000 y BP.

Over large areas of central London, Silt Complex sediments are known to have occurred but have been exploited over a long period for brick-making (Bromehead, 1925) and are therefore absent or of very reduced thickness. The borehole cross sections in the London area confirm this situation. For example, in the Holborn–Whitechapel section (Fig. 26) 82 cm of 'brown clayey sand' rests on Corbets Tey Gravel in borehole 682 and 2.0 m of 'yellow clay' overlie Mucking Gravel beneath St Paul's (borehole 30). Further east, typical 'brown silty clay with stones', 2.26 m thick is present east of Liverpool Street. In Tower Hamlets almost all the Silt Complex sediments are absent except in borehole 59 where 80 cm of 'brown silty sand with some gravel' occurs.

This situation is found in most sections from central London (for example, Figs 10, 16, 17, 18, 19, 20, 35, 41). However, in some areas more extensive spreads occur. As already mentioned (section 2.7j), sediments described as 'brickearth' are present beneath much of north-east London, particularly in the Hackney–Stoke Newington–Highbury areas. At Highbury, thicknesses of up to 12.25 m were reported by Tylor (1868) and are confirmed by recent boreholes (Figs. 18, 27). This sediment is described as 'alternating bands of sand and clay, grey to brown in colour' or 'brown silty clay with gravel' in these holes and, in view of the position of this spread, adjacent to the London Clay Highbury hill, it is almost certain that these sediments represent colluvial material soliflucted from the hilltop forming a hillside fan. The sequence is precisely similar to that described from Northwold Road–Stoke Newington Common area (Harding & Gibbard, 1984). In the latter area, similar sediment rests directly on fluvial sands and gravels thought by these authors to be of Late Devensian age.

The deposits occur in a similar situation south of the river. They are particularly thick at Peckham where they overlie the fossiliferous sediments (section 2.7l). This 'brown clayey sand' or 'pebbly sand' reaches to 6.25 m in thickness against the steeply rising London Clay to the south. This position suggests an origin comparable to that at Highbury. The sediments continue as a veneer on the Kempton Park Gravel to the north. They were examined in the field in a gas pipe trench in New Cross (TQ 353762) where they comprised 1 m of orange-brown silty sand with isolated pebbles.

On the Lea–Roding interfluve, Silt Complex sediments seem to be lacking, with the exception of the Upton Park area (TQ 420834: 7 m) where boreholes confirm that 1.75 m of these deposits cap the local gravel and sands. Whitaker (1889) also recorded 'brickearth' sediment in old workings adjacent to the Northern Outflow Sewer at Plaistow (TQ 406826: 6 m).

Immediately east of the Roding, the large mass of 'brickearth' sediments at Ilford occur. These sediments, already described in detail above (section 2.7i; Fig. 43), cloak the underlying deposits and cause the problem noted by Dines & Edmunds (1925, p. 34): 'in a large tract of river deposits that occupies the south-western portion of the map they merge into one another.' Indeed, detailed investigation demonstrates that the 'brickearth' here rests not only on pre-existing fluvial sediments, but also on London Clay bedrock. Brown mottled grey silty clay sediment, 1 m thick, unconformably overlies the gravel and sand at Fairlop (TQ 451895: 28 m) (Fig. 13).

Discontinuous patches of 'brickearth' sediments are encountered towards the east, but the next most continuous spread is found in the Purfleet–Ockendon–Corbets Tey 'meander' area. Here a thin veneer (50 cm) of brown to mottled grey-brown pebbly clayey silt occurs, cryoturbated throughout. In places, for example at the Corbets Tey, Bush Farm type section (TQ 570844: 22 m), the upper 1–1.5 m of the gravel and sand beneath the brickearth' is highly oxidised, is orange in colour and is itself cryoturbated. The latter appears to predate the emplacement of the 'brickearth', and may therefore indicate a substantial period of subaerial exposure of the deposits beneath, before arrival of the overlying sediment. Silt Complex sediments are represented in the M25 section (Fig. 12) where they reach up to 9 m thick in the North Ockendon area. At Purfleet (TQ 568785–570797) brown sandy silt with isolated pebbles, 30 cm to 3 m thick, occurs again, backed against the chalk slope immediately to the south. This position suggests a colluvial origin (section 2.5). This sediment is 1.75 m thick at Marley's Pit, Aveley (TQ 572802: 17 m).

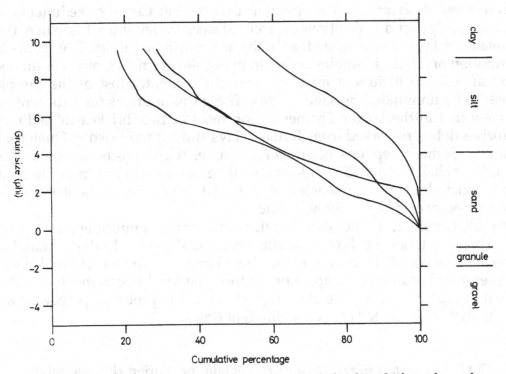

Fig. 43. Particle-size distribution of samples of clayey silt 'brickearth' from the underpass section, Riches Road, Ilford.

The complex sediment sequence exposed in South Stifford Quarry has already been mentioned (section 2.5). Here Corbets Tey Gravel abuts an interbedded sand and silt unit that is capped by brownish-red massive sandy silt 70–150 cm thick. Brown sand clay with gravel up to 1.9 m thick caps the sections at Aveley, Sandy Lane (TQ 552807: 14 m). Dark grey-brown sandy clay with pebbles and broken Woolwich Beds shells overlie the Crayford silts (e.g. TQ 517758) and brown clayey sandy gravel 2 m thick overlies the sediments at West Thurrock (TQ 598782: 17 m and Fig. 31) and 1.8 m of silty sand at Grays Thurrock (TQ 623781: 12 m). 'Brickearth' has been described from the Low Street area by Dewey et al. (1924). They comment 'in a shallow railway cutting half a mile north of Low Street station sandy brickearth can be seen, while east of the station there is a disused brick-pit where sandy loam with very few stones is exposed.'

The brown silt overlying the Dartford Heath Gravel (section 2.2) at Bowman's Lodge (TQ 518738), and filling a channel from the indistinct stratification and variable thickness, appears to represent a slope wash or mass flow deposit possibly derived from London Clay. In the Swanscombe area, 'brickearth' sediments occur in Knockhall Cutting (TQ 595745: 19 m). Here a small valley cut into the famous Swanscombe sequence (section 2.3) is infilled by 2.5–3 m of orange buff clayey silt with occasional stones resting on weathered chalk. It was well exposed immediately north of the footbridge. The deposit was cut out to the north by the rising chalk floor and showed the typical vertical columnar joint drying structures. This material was very similar in appearance to that described from Northfleet by Kerney & Sieveking (1977) and Kemp (1991) (section 2.7a).

From these descriptions it is apparent that the Silt Complex sediments are predominantly found in positions adjacent to steep valley sides. Elsewhere they are relatively thin or absent, such as in the area north of Tilbury. The grain-size distribution of selected samples is shown in Fig. 44. From this, much of the sediment appears to include a relatively low silt content. Most of the samples appear to be dominated by sand or clay. These distributions tend to reinforce the view that in the Lower Thames region most of this 'brickearth' sediment comprises debris reworked from Tertiary clays and/or pre-existing Quaternary sediment. Some exceptions occur: for example, the deposits at Ilford, Bush Farm, Knockhall Cutting, Seven Kings, Collingwood Farm and possibly South Stifford include high frequencies of silt that might be of aeolian origin. However, other origins cannot be excluded.

Overall, therefore, it is obvious that this sediment is of multiple origin, much of it representing local colluvial (solifluction and slope wash) deposition. Part of the material could also represent weathered and partially remobilised estuarine interglacial sediments in appropriate topographical situations. Its conflicting stratigraphical occurrence also suggests that it is probably polycyclic and potentially has originated at several different times.

2.11 Shepperton Gravel ('Lower Floodplain' or 'buried channel' infill)

Dewey & Bromehead (1921) introduced the term 'Lower Floodplain' gravel to separate the lowermost gravel and sand unit in the Thames system from the higher or 'Upper Floodplain' gravel (Kempton Park Gravel; section 2.9). This 'Lower Floodplain' gravel was thought by Dines *et al.* (1924) to be continuous

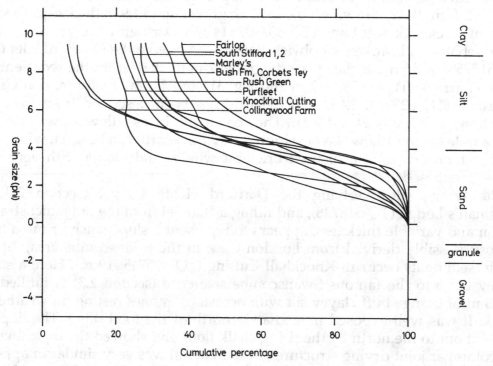

Fig. 44. Particle-size distribution of samples of clayey silt 'brickearth' from various sites mentioned in the text.

2.11 SHEPPERTON GRAVEL

with the deposits filling the 'buried channel'. The latter was the name given to the infilled valley cut into bedrock and grading downstream beneath the modern floodplain eastwards from the London area. King & Oakley (1936) recognised that the 'Lower Floodplain Terrace' gravel (their Halling Stage) was 'co-extensive with the main Buried Channel of the Lower Thames Valley' and consists 'of gravels which protrude through the recent alluvium in the neighbourhood of London as 'islands' (eyots).' They concluded therefore that the filling was a 'normal fluviatile aggradation'.

In the Lower Thames area the modern floodplain has received relatively little attention from geologists. This is mainly because sections are extremely rare; excavations being invariably flooded (cf. Gibbard, 1985). Notable exceptions are the informative review by Spurrell (1889) and the London Memoir by Whitaker (1889). The former author gives a detailed account of the gravel and sand, recognising that it is continuous beneath the modern valley from Richmond to Gravesend and beyond. More recently, some observations, particularly of tributary floodplain gravel, were made by the geological surveyors (Dewey et al., 1924; Dines & Edmunds, 1925). Most of their information for the Thames itself was based on borehole records and much of the sediment assigned to their floodplain gravel is equivalent to older units, as discussed in sections above.

Following previous authors, particularly Hare (1947) and Briggs & Gilbertson (1980), Gibbard (1985) suggested that, since the floodplain gravel was the most recent sand and gravel aggradation (i.e. it is undissected and unmodified since deposition), it could serve as an analogue for the interpretation of the older gravel members. This, he felt, justified detailed examination of the unit. Moreover, for stratigraphical consistency it should be formally defined as the other lithostratigraphical units and separated from the overlying lithologically contrasting, predominantly fine, floodplain sediments (Staines Alluvial Complex/Tilbury Deposits, section 2.13). The term 'Shepperton Gravel' was selected for the sand and gravel member, defined from Shepperton Quarry (TQ 070669: 11 m) where the deposits were well exposed (Gibbard (1985, p. 70). Since there is a complete lack of exposure of the floodplain gravel unit in the Lower Thames area, a complete investigation of its lithology and sedimentary structures was not possible. However, the unit is a direct continuation of that in the Middle Thames; therefore no new lithostratigraphical unit is defined, and instead the term 'Shepperton Gravel' will also be used to refer to the unit in this area.

Evidence that the deposits in the Lower Thames area are indeed a direct continuation of those upstream can be clearly demonstrated in central London, both north and south of the river (see Fig. 48 below). In the Southwark area, the floodplain is 2 km wide and is underlain by 5–6 m of gravel and sand beneath 1–3 m of organic sediment and 1–2.5 m of fill (Gibbard, 1985, fig. 43). This same unit was followed beneath the river to the north bank beneath Whitehall and the Embankment and occurs in the section from beneath Blackfriars to Borough (Fig. 40). On the south side of the river the Shepperton Gravel abuts the East Tilbury Marshes/Kempton Park Member (section 2.9) north of the Elephant & Castle. This junction can be traced eastwards using borehole records towards Bermondsey beneath the London Bridge–Deptford railway line. Likewise, the sequence is repeated beneath London Bridge, at the north end of which the unit abuts the Mucking/Taplow Member (Fig. 34).

Berry (1979) describes a shallow enclosed bedrock hollow beneath the northern end of Blackfriars Bridge (TQ 316808). This elongated depression is about 200 m in length and reaches a maximum depth of –9.7 m OD, i.e. *c.* 2 m below the 'normal' base of the Shepperton Member in the area. It is infilled with sand and gravel. This feature appears to be associated with the confluence of the River Fleet, although Berry says that similar hollows are common all along this Kings Reach area of the valley, adjacent to the valley side. Other shallow examples occur beneath London Bridge, where the Walbrook stream joins the Thames (TQ 327806), and Hay's Wharf (TQ 331803).

The Shepperton Gravel is poorly represented north of the river as far east as Limehouse Reach, apart from narrow spread at the London Docks, Shadwell (TQ 352806). At this site 4.28 m of gravel and sand underlie the alluvium (Whitaker, 1889), but thicknesses of up to 6 m were reported by Bromehead (1925). However, to the south the spread widens considerably to 3.25 km south and east of Bermondsey, equivalent deposits occurring as far south as New Cross Hospital (TQ 353775). This area was the valley of a 'lost river', the Earl's Sluice from Camberwell and its tributary, the Peck from Peckham Rye (Barton, 1962). Boreholes at Surrey Docks record gravel and sand beneath organic deposits (Fig. 1) at the appropriate height for correlation with this unit (cf. Dewey & Bromehead, 1921; Whitaker, 1889). Immediately east, within the substantial Greenwich meander, is the Isle of Dogs. Here the entire area, within the meander as far north as the East India Dock Road, is underlain by floodplain gravel and sand (Fig. 41). This was seen in excavations for Millwall Docks (TQ 3779) in 1866 (Whitaker, 1889) and the West India Docks in 1854 (Blandford, 1854). At the eastern end of this spread, the Rivers Thames and Lea floodplains merge between Bromley and Canning Town. Here it is also possible to demonstrate the continuity of the floodplain gravels beneath the two valleys. For example, at Poplar Fire Station (TQ 381812: 5.5 m) 5.04 m of gravel and sand underlie 5 m of made ground. The deposits are next encountered in boreholes for the Blackwell Tunnel northern approach road (Fig. 1) where they reach 6.41 m thick and are overlain locally by brown silty clay (?Tilbury Deposits). Beneath the north side of the modern channel at Blackwall another example of an elongate scour-type hollow was reported by Berry (1979). This gravel and fine sediment-filled depression reaches a maximum depth of –30.5 m above the western tunnel bore (TQ 385802). A possible continuation of this hollow occurs at Trinity Wharf at the Thames–Lea confluence.

South of the meander at Deptford the Ravensbourne joins the Thames. Immediately to the east, the modern Thames is banked against East Tilbury Marshes Gravel (section 2.9) beneath Greenwich, as far east as the Blackwall Tunnel southern approach road (Fig. 42). Beyond New Charlton as far as Erith, the south margin of the floodplain is a steep bedrock slope (Figs. 45, 46).

East of the Lea confluence, the floodplain is a continuous strip 1.5–2 km wide, as far as Purfleet. Excavations adjacent to the confluence, for the Royal Victoria (TQ 4180) and Royal Albert Docks (TQ 4280) were reported by Blandford (1854) and Whitaker (1889) respectively. Both observed gravel and sand beneath fine floodplain sediments; at the former, 6 m were seen resting on London Clay but at the second only 1 m was seen (unbottomed). The local sequence beneath the floodplain is shown in Figs. 38 and 39.

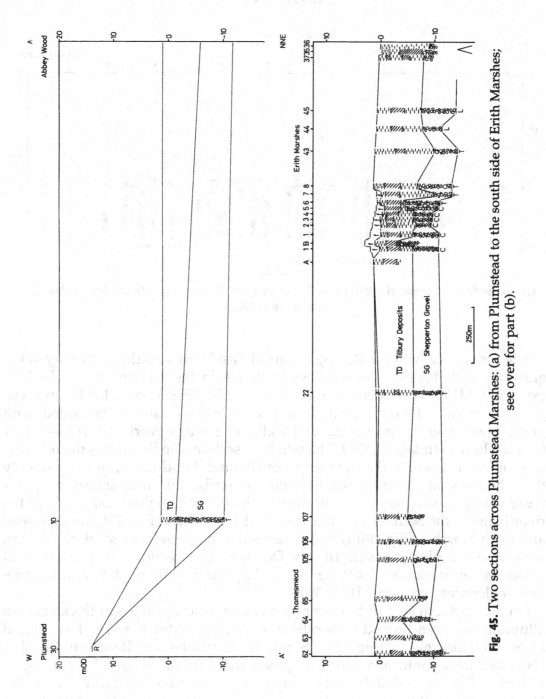

Fig. 45. Two sections across Plumstead Marshes: (a) from Plumstead to the south side of Erith Marshes; see over for part (b).

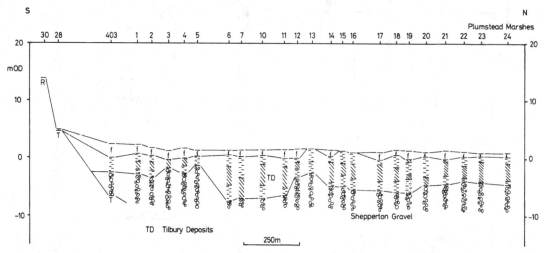

Fig. 45. (b) from Plumstead north to the Southern Outfall Sewer, constructed using borehole data (source: GLC).

The confluence with the Roding occurs at East Ham, and the continuity of the gravel spreads beneath both valleys is shown in the sections along the M11 extension, M15 and the Becton river crossing (Fig. 46; section 2.12c below). The gravel and sand ('coarse sand, locally false-bedded, and interbedded with gravel') was seen in an exposure at Gallions Sewage Works (c. TQ 4481) by Goodchild (in Whitaker, 1889). The borehole section also illustrates the relationship of the deposits to the modern river channel. In addition, two or possibly three hollows, of a similar scale to those described from upstream by Berry (1979), are present. Here the continuity of the deposits on the south side of the river beneath the Southern Outfall Sewer at Plumstead (Fig. 45), Thamesmead and Erith Marshes (Fig. 46) can also be seen. The gravel and sand of this unit were also recorded beneath Tilbury Deposits in a section for a reservoir at Crossness by Whitaker (1889, fig. 98) and Spurrell (1889, p. 218) and in boreholes at Jenningtree Point (TQ 503803).

On the northern side of the river, gravel and sand up to 6.5 m thick beneath alluvium has been proved in boreholes on Barking Marshes (TQ 476825) and at South Hornchurch (Fig. 24). Similarly, boreholes on Rainham Marshes (TQ 5380) have confirmed over 2 m gravel and sand of this unit beneath up to 11.6 m of fine alluvial sediments near the river (cf. Dines & Edmunds, 1925). However, the unit thins towards the eastern valley side and is only 1–1.25 m thick at TQ 530800 (0.6 m OD). At Purfleet the floodplain is constricted by the rise of the Chalk bedrock Purfleet anticline; indeed the gravels thin markedly against this high, such that beneath Aveley Marshes the Shepperton unit is only c. 1 m thick (e.g. TQ 535791). A similar thinning and constriction occurs on the adjacent south bank at Erith.

Downstream of Erith lies the compound confluence of the Darenth and Cray streams. Once again the gravel spreads beneath these tributary valleys appear to merge with that of the main river. Beneath Dartford Marshes 2 m of gravel and sand of the Shepperton Member occur beneath the Long Reach Hospital (TQ 552772; Dewey et al., 1924). Boreholes for installations west of the Dartford Tunnel approach record up to 8.55 m of gravel and sand beneath 9 m of allu-

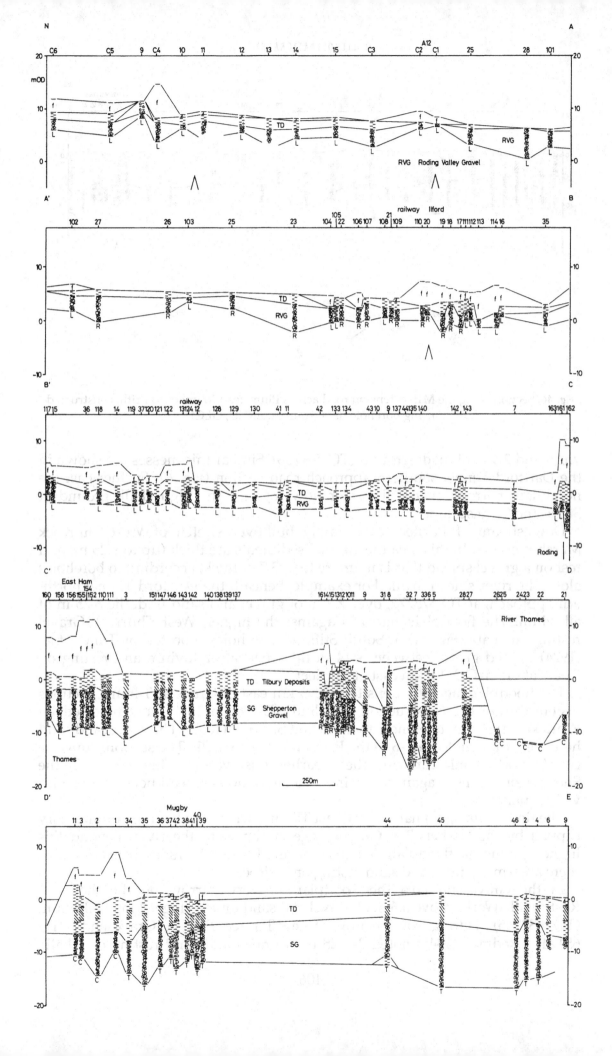

2 LITHOSTRATIGRAPHY

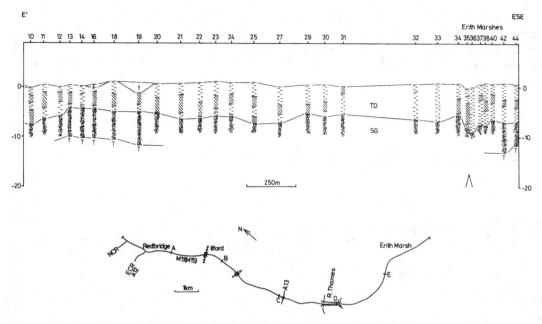

Fig. 46. Section along the M15 extension road across Plumstead Marshes to Erith, constructed using borehole data (sources: ERCU, GLC).

vium and 2.4 m of made ground (TQ 561758). Similar thicknesses are shown in the Dartford Tunnel southern approach road section (Fig. 6). Further east the floodplain narrows where the river flows against the high chalk ground at Stone and Swanscombe.

Downstream of Purfleet the heavily builtover stretch of West Thurrock Marshes occurs. In this area the alluvial sediments are thick (up to 9.25 m) and rest on a gravel spread that is no more than 3.7 m thick, according to boreholes along the river's north bank. For example, beneath the Dartford Tunnel northern approach, at TQ 575772, over 2.3 m of gravel and sand underlie 8.25 m of alluvium. The floodplain narrows against the higher, West Thurrock Gravel resting on chalk (Fig. 31) at South Stifford. Boreholes reported by Dines *et al.* (1924) proved gravel between 6–14 m thick beneath alluvium and resting on Chalk at –15 m OD, 1.6 km east of Purfleet.

The floodplain again widens to over 3 km east of Grays at Tilbury. Sections visible during the construction of Tilbury Docks in 1885 were reported by Whitaker (1889 and references therein) and Spurrell (1889). The gravel and sand here were thought to be 3–6 m thick and rested on Chalk. The section along the river's north bank confirms these earlier observations (Figs. 36, 47), the Shepperton Member again occurring as a continuous spread beneath the alluvial sequence.

The floodplain again narrows at East Tilbury, where the bedrock promontory capped by the Corbets Tey Gravel projects almost to the river. Beyond this higher ground the floodplain widens to form Mucking Marshes, beneath which there are 8 m of gravel and sand resting on bedrock.

In the Gravesend area the floodplain is very narrow but boreholes at Rosherville Works prove 6.6 m of gravel and sand underlying 9.4 m of alluvium (TQ 633745). During construction of the Terrace Pier at Gravesend (?TQ 651744), Redman (1845) noted 3–4.25 m of gravel and sand resting on Chalk.

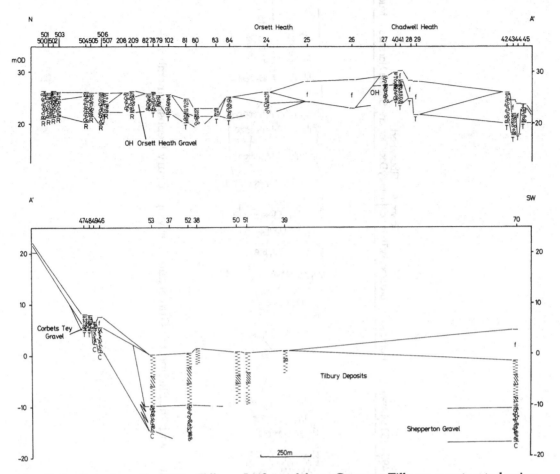

Fig. 47. Section along the A13 – Tilbury Link road from Orsett to Tilbury, constructed using borehole data (sources: ECC, ECRU).

Below the town to the north of Allhallows there is a continuous marsh area. Beneath Shorne Marshes the deposits have been proved in a borehole south of Shornmead Fort (TQ 692746). This recorded 9.2 m of gravel and sand beneath 10.4 m of alluvium.

2.12 Tributary valley floodplain gravels

Several important tributaries join the River Thames along its course from central London to Tilbury. In most cases these tributary valleys are floored by gravel and sand aggradations that can be correlated with, and can frequently be shown to grade into, the Shepperton Gravel (section 2.11; Fig. 48). The implications of these correlations are very important for relating the evolution of tributary valleys to the Thames and vice versa. The tributaries for which sufficient information was available during the study will be dealt with in order downstream from central London.

(a) River Fleet

The River Fleet is one of the most famous 'lost rivers' of London. According to Barton (1962), it rose on Hampstead Heath, then flowed south to Camden

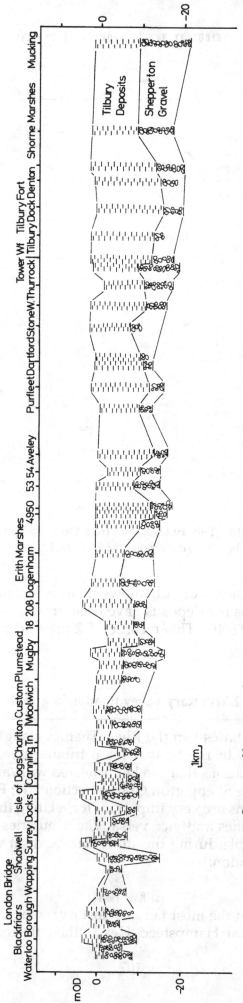

Fig. 48. Long profile section showing the deposits correlated with the Shepperton Gravel and overlying Tilbury Deposits members.

Town, to near King's Cross, along the line of Farringdon Road, beneath Holborn Viaduct and entered the Thames immediately west of Blackfriars Bridge. The valley is shown on the Geological Survey sheet 256 as being underlain by alluvium. This course dissects the Thames deposits on the north side of the river (Fig. 2). Moreover, a minor tributary of the Fleet, which rose in north Bloomsbury immediately south of Euston Road, may have been that responsible for deposition of the Endsleigh Gardens channel sediments described by Gibbard (1985, p. 60).

The Fleet Valley is shown in four borehole sections (Figs. 19, 20, 26, 40). The most southerly of these prove gravel and sand deposits 2.14 m thick beneath floodplain peat and fill, immediately north of Blackfriars Bridge. The north-south section shows the continuity with the Shepperton Gravel. A similar thickness and stratigraphy occurs beneath Holborn Viaduct, where the Lynch Hill/Corbets Tey Gravel is dissected. Likewise, the section along Clerkenwell Road again shows the narrow Fleet Valley cut into the same Thames unit.

According to Berry (1979), an enclosed hollow underlies part of the Fleet Valley beneath Calthorpe Street (TQ 310824). It is aligned NW-SE, and is about 305 m long and 6 m deep. Berry concludes that it is probably did not originate beneath the Fleet, but beneath the older gravel spread (Lynch Hill/Corbets Tey Gravel) that caps this area. The Fleet may have cut into the eastern end of the feature.

(b) River Lea

The River Lea (or Lee) is the largest of the tributary streams in the Lower Thames area and its modern floodplain is seldom less than 0.8 km wide in London. It is confluent with the Thames at Canning Town. Geological research on the Lea has a long history. The earliest detailed investigations were carried out over 50 years by S. H. Warren. In a series of papers (Warren, 1912, 1916, 1923, 1940; Allison, Godwin & Warren, 1952) he described the occurrence of organic, plant-bearing beds interstratified with and overlying gravel and sand units beneath the floodplain in the lower Lea Valley. These beds contained cold climate or 'full glacial' plant assemblages and were therefore referred to collectively as the Lea Valley Arctic Bed. Warren stated that 'if it be desired...to give a name to this stage of the Pleistocene, I would suggest that the term 'Ponder's End Stage' would be an excellent one' (1912, p. 219). This name, being derived from the locality from where the bed was first discovered, was applied not only to the plant beds, but also to the gravels in which they occurred.

Although the beds were initially found in the gravel pit at Pickett's Lock (TQ 3595) near Ponder's End, later discoveries included sites at Edmonton (Angel Road, Hedge Lane and later Barrowell Green), Temple Mills and Hackney Wick. His detailed reports contain valuable descriptions of these localities that have all long since been infilled, flooded or built over. Unfortunately, it is beyond the scope of this work to discuss the geology of the Lea Valley in detail. However, it is clearly of considerable importance to establish the relationship of the Lea Valley gravel units to those of the Thames for purposes of establishing the contemporary environment and possible dating.

At the Pickett's Lock a sequence of 5.5 m of stratified gravel and sand beds rested on London Clay and was capped by 1.2 m of 'brickearth'. The section drawing clearly illustrates that the gravel and sand were disposed into current bedded units similar to those described from high level units in the Thames and tributary valleys (Chapter 3). The dark blue-black organic bed occurred over most of the area of the workings. This bed varied considerably in texture from a clay to a sand and with varying contents of plant detritus, beetles, shells and occasional bones. Warren (1912) records clearly that the bed was considerably disturbed in places, occurring as raft-like detached blocks. He particularly notes one point where 'three large masses of Arctic Bed were seen in vertical succession, separated one from the other by strata of gravel and sand', these stacked blocks clearly having been picked up with the gravel and deposited in a braid-like accumulation.

A radiocarbon date of 28 000 ± 1500 BP (Q-25) was obtained for fine silts in gravels of the lowest terrace of the Lea by Godwin & Willis (1962) and 21 530 ± 480 BP by Shotton & Williams (1971). However, it is unclear from which site the sample was taken; possibly Nazeing or Broxbourne, Essex. Nevertheless, deposits of this age have been confirmed from the Broxbourne area by A. Lister who obtained a date of 24 630 + 1360–1640 BP (Q-2832) on an elephant tusk from Ryhoff Quarry (Lister & Switsur, unpublished).

Similar detailed descriptions are given for other sites and from these it is apparent that the sediments containing the 'Arctic Bed' were extensive beneath a 'Low Terrace' 1–2 m above the modern floodplain. This 'terrace' extends from Broxbourne to Tottenham, on the western side of the river (Whitaker, 1889). However, from the descriptions, it is clear that the underlying deposits are continuous, at least in part, with the sediments beneath the alluvium. Indeed, Warren (1916) concluded of the sediment at Hackney Wick that 'there is little doubt that it should be grouped with the buried channel Drift'. He also believed that 'the buried channel must be cut through the Low Terrace' deposits. If this is the case then it appears that the gravel and sand accumulation of the Ponder's End Stage must predate the final 'buried channel' gravel unit (cf. Wymer, 1985). However, neither the previous descriptions nor modern observations offer clear evidence for an erosional event between these two accumulations. Therefore, until the matter can be further clarified, the noncommital member term 'Lea Valley Gravel' is proposed here for the gravel and sand unit underlying the modern floodplain and Warren's Low Terrace. The Ponder's End section is proposed as the stratotype. If a separate aggradational phase is subsequently identified, however, the term 'Ponder's End Gravel' could be revived for the older unit.

From the publications it is not entirely clear whether Warren considered his various finds of organic sediment in the gravels as indicating broadly a single discrete bed that had been later modified by fluvial activity, or a complex of individual beds. Experience suggests that it is unlikely that the material at all the sites could represent a single aggradation, but could be a series of channel or wider scale depression fills (Chapter 3). If this is so then the 'Low Terrace' may simply represent an accumulation in the wider part of the valley that escaped later modification. Whatever the situation, the Lea Valley sequence deserves detailed modern investigation, since in this report only the London area sections are described (Fig. 1).

As noted above, the Lea and Thames are confluent at Canning Town, where boreholes confirm the 2.44 m of gravels and sands beneath 6.88 m of floodplain alluvium (TQ 391814: 4.4 m). Upstream at Mill Meads, West Ham, a series of boreholes prove gravel and sand up to 6 m thick beneath the alluvium: for example at TQ 387828 (5.5 m). Similarly, the boreholes for the Northern Outfall Sewer demonstrate that the unit continues northwards to beneath the Eastern Region main line at Stratford Marsh. Boreholes at the western end of Stratford High Street also confirm that the member is present beneath the floodplain.

A transverse section across the valley along the East Cross–Hackney–M11 link route (Eastway–Ruckholt Road) is shown in Fig. 14. Here the floodplain extends from Mabley Green, Homerton, to beneath Temple Mills marshalling yard, a width of 1.9 km. In this section the Lea Valley Gravel is of rather irregular thickness (up to 6 m), shows an undulating base and upper surface and is everywhere overlain by alluvial sediment. The member rests on Tertiary bedrock. According to Warren's (1916) location map, the Hackney Wick site, from which he recovered peat and vertebrate material from the gravel unit, was approximately where the Arena Field Stadium now stands (c. TQ 373850: 4.6 m).

The unit is next encountered under Lea Bridge Road where it is again proven by boreholes at TQ 361870 (6 m) to be 3.2 m thick and is overlain by 2.9 m of brown clay and made ground.

The second cross-section of the valley is along the Victoria Line (London Underground) that runs under Forest Road and westwards beneath the railway line to South Tottenham (Fig. 15). As in the East Cross section, the floodplain forms a distinct feature, 1.8 km wide, underlain by the Lea Valley Gravel Member. The gravel spread is again of irregular thickness, both the base and surface showing distinct undulations. On the eastern side, the gravel member is clearly thicker, reaching a maximum of 6.5 m. This locally thicker accumulation may be analogous to the 'Low Terrace' of Warren (1916), although he did not recognise the feature on this side of the valley. By contrast, he did map the 'terrace' beneath Tottenham. The borehole section (Fig. 49) along the so-called North–South Route through Tottenham crosses Warren's mapped boundary of the floodplain and his 'Low Terrace' but shows no apparent change of level of the units, suggesting possible continuity. It should, however, be stressed that this does not proclude the possibilty that there are two separate accumulations, at approximately the same altitude, the gravel aggradation beneath the flood-

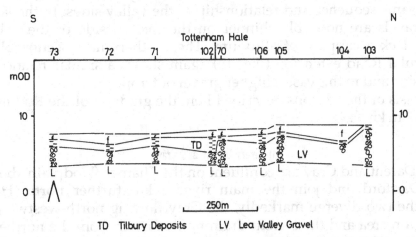

Fig. 49. Section beneath Tottenham, constructed using borehole data (source: GLC).

plain having been deposited in channels excavated into the Ponders End accumulation (as Warren contended).

Evidence in support of the latter conclusion is that the gravel unit beneath the 'Low Terrace' is partially overlain by 'brickearth', rather than floodplain alluvial sediments (Whitaker, 1889; Warren, 1912; 1918; Dewey & Bromehead, 1921). This might suggest a greater antiquity for the underlying gravel and sand.

The Lea Valley Gravel can be traced upstream at least as far as the Hertford area (cf. Gibbard, 1974, 1977). Plotting of the sediment thicknesses indicates a downstream gradient of 75–95 cm km^{-1} (Fig. 22). On the basis of counts by Warren (1916), the gravel comprises up to 95% flint (70–90% angular flint, 15–45% rounded flint), 1% vein quartz, 1–5% quartzite and 0–1% Greensand chert and 1% igneous and metamorphic rocks.

(c) River Roding

The River Roding drains southern Essex, flowing south-westwards as far as Buckhurst Hill and then turning SSE from Wanstead to meet the Thames at Barking Creek, at East Ham. The lower part of the valley cuts through a series of Thames' terrace sediments resting on London Clay throughout the area considered here. The valley deposits of this stream have not been studied in detail, but are mentioned in passing by Whitaker (1889) and Dines & Edmunds (1925). As already mentioned, there is a high density of boreholes in the East Ham–Barking area associated with sewerage work and construction for the M11 and M15 motorways (Fig. 46). From these it can be seen that a unit of sand and gravel floors the Roding Valley and is overlain by alluvial sediments. These units merge laterally with their Thames' equivalents. The 8 km long section along the M11 from South Woodford to East Ham demonstrates that this Roding Valley Gravel unit is continuously present beneath the floodplain. The unit varies in thickness from 2 m to 5 m and at one locality includes 80 cm of brown-grey silty clay with shells (borehole 14). The upper surface of the gravel unit is also undulating and shows an amplitude of 1–1.5 m.

Three sections across the valley have been reconstructed from borehole records. The first at Ilford (Fig. 25) illustrates the relationship of the valley gravel spread to the deposits forming the valley sides, in particular the fossiliferous sediments discussed above (section 2.7i). At this point the valley is 300 m wide at Ilford Hill. However, at Redbridge the valley is almost 500 m wide, but shows the same sequence and relationship to the valley sides. In the latter section the gravels are noticeably thinner on the eastern side of the valley and apparently lack a capping of alluvium. The northernmost section along the North Circular Road extension (Fig. 15) again shows a similar relationship to the valley side and in this case a higher gravel outcrop.

On the basis of the sections described here the gradient of the Roding Valley Gravel is 1.2 m km^{-1}.

(d) Rivers Darent and Cray

The rivers Darent and Cray are confluent on the Thames floodplain about 1 km north of Dartford and join the main river 2 km further north. However, upstream the two diverge markedly, the Cray flowing north-westwards from the Orpington area and the Darent draining due north from the northern edge of the Weald at Sevenoaks. No sections or borehole records were available for

2.12 TRIBUTARY VALLEY FLOODPLAIN GRAVELS

the Cray Valley during this study. However, some information was found for the Darent.

A section across the Darent Valley along the M25 motorway is shown in Fig. 50. This shows an apparent twofold division of gravel and sand aggradations beneath the valley. The higher, older unit underlies the western valley side beneath the A225 road. This unit is about 2.75 m thick and is overlain by 2.5 m of 'brown sandy clay and flints'. For convenience, the term 'Hawley Gravel' is proposed for this unit (type section: M25 at Hawley). The geological map (sheet 271) shows this unit as a continuous sheet from Sutton-at-Hone downstream to Dartford. It is not known whether or not this represents a single unit. However, the lower of the two spreads in the M25 section is clearly separable on the basis of altitude.

This lower spread of sediments was exposed in an adjacent gravel quarry at Sutton Place in 1986 (TQ 555714: 13 m). Here 3 m of deposits were exposed resting on an irregular chalk bedrock surface. The sediments comprised predominantly horizontally bedded medium gravel with thin sand interbeds. Large sarsens up to 45 cm in diameter were common at the base of the gravels. However, at two places in the pit, substantial channel-like fills of light grey to brown-grey silty clay to silt occurred. The first unit of this fine sediment was 1.4 m thick, the base resting conformably on thin basal gravel. In the upper 70 cm the unit contained thin sand bands 5–8 cm thick that were much disturbed by loading. The upper surface was truncated by an overlying bed of coarse gravel 1.4 m thick. A second unit of fine sediment 95 cm thick occurred some 50 m to the north. The upper 10–15 cm of the silty clay contained frequent isolated large stones. Both these units are laterally impersistent. The section is capped by floodplain alluvial black organic detritus mud up to 1.25 m thick on the south side of the site. The gravels almost totally comprise flint, with less than 2% of small quartz pebbles and Greensand chert. This section is proposed as the stratotype for these Darent Valley Gravels.

Previous descriptions are rare. Dewey *et al.* (1924) note that gravels and sands were found near Hawley Mill (TQ 553718: 12 m) and beneath Dartford

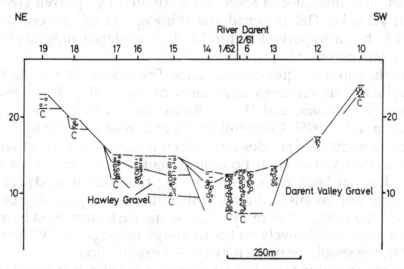

Fig. 50. Section across the Darent Valley along the M25 motorway, constructed using borehole data (source: ERCU).

(*c* TQ 546728: 7.6 m) where the sediments were *c*. 6 m thick. Whitaker (1889) describes a series of gravel exposures in the vicinity of Dartford, particularly beneath the Dartford–Hawley road. However, as will be apparent from the comments above on the M25 section, at least some of these exposures were possibly in the higher, Hawley Gravel rather than the valley bottom Darent Valley Gravel.

The lack of information prevents direct correlation of these Darent Valley Gravels with the Thames sediments. However, the similarity of their occurrence suggests that they are probably confluent downstream with the Shepperton Member, as in other valleys.

2.13 Staines Alluvial Deposits/Tilbury Deposits

Alluvium is the term that has been applied to the predominantly fine-grained deposits of silt, clay, mud, peat, marl, tufa, sand and gravel that underlie the marsh and water meadow land fringing the rivers in southern Britain. These floodplain areas are liable to flooding when the rivers overflow their banks, although human interference (locks, dykes, weirs, etc.) has greatly reduced the frequency of such events.

The modern floodplain of the River Thames is variable in width. According to Whitaker (1889), upstream of London the alluvial deposits occur as narrow strips bordering the river (cf. Gibbard, 1985), whereas downstream there are broad marshes on either side. Whitaker was in no doubt that 'this fact had much to do with the selection of the site of London, which is built on the first...broad low-lying tract of dry land.' Under natural conditions the marshes would be inundated at high tide since they are below high tide level. However, the construction of embankments has prevented flooding of these areas. Presumably the marshes would have looked like the large saltings and mud banks that occur now in the estuaries of some of the south Essex rivers. This flood protection has resulted in cessation of mud accumulation on the marshes, so that their surface now occurs at a lower surface level than that of modern mud banks. The difference in level was attributed by Spurrell (1885) to subsidence, but Whitaker (1889) considered shinkage resulting from drying of the sediments to be an important factor. Modern evidence suggests both may be contributing to this effect.

Work on the alluvial sequence both in the Thames and its tributaries has been limited. Principal descriptions have been by Spurrell (1885, 1889), Whitaker (1889), Dewey & Bromehead (1921), Bromehead (1925), Dewey *et al.* (1924), Dines & Edmunds (1925), Churchill (1965) and most recently by Devoy (1979). In addition, a number of studies have taken place in connection with archaeological excavations, mainly in London. Detailed investigation of the alluvial sediments has not been undertaken during the present study. However, for completeness and because alluvial sediments have repeatedly been encountered during the work, a summary of the sequence is presented here. This summary relies almost exclusively on the thorough monograph by Devoy (1979), to which reference should be made for detailed explanation.

The records available from the numerous boreholes that penetrate the alluvial sequence are not satisfactory for tracing individual beds. This is because the boreholes were not recorded with sufficient care and the necessary consistency to distinguish the internal variability of the alluvial sediments. In many cases beds were recognised but it is not possible to be certain to which of Devoy's units they relate. For this reason no attempt will be made here to correlate the beds from borehole records.

Devoy (1979) investigated local and regional sequences in the Lower Thames valley between Central London and the Isle of Grain. He recognised five discrete biogenic beds (numbered Tilbury TI–V upwards), separated by inorganic beds (Thames I–V upwards) based on a stratotype at World's End, Tilbury (TQ 64667540: 2 m). Because Devoy did not propose a formal name for the whole alluvial sedimentary unit, the term 'Tilbury Deposits Member' is used here for the predominantly fine-grained sediments overlying the Shepperton Gravel at the type section and throughout the area. These deposits pass laterally into the Staines Alluvial Deposits as defined by Gibbard (1985) for the Middle Thames area. This is in complete agreement with the previous observations by Spurrell (1889, p. 212).

According to Devoy, the gravel surface at the Tilbury type locality generally occurs at –9.5 to –10 m OD, but shows some clear channel-like depressions and highs, the gross amplitude being locally almost 8 m. Five biogenic beds interbedded with olive-green to blue-grey silt and or clay beds occur. The biogenic sediments show much internal facies variabiltity, both laterally and vertically, apparently reflecting changes in depositional conditions. In contrast, the intervening inorganic units are uniform in character. The basal biogenic unit (Tilbury I) occurs at between –12 and –16.5 m OD and comprises a black gyttja or wood peat 15 cm thick. Tilbury II is represented by dark brown wood peat that changes upwards into a sedge peat and occurs at an average depth of –10 to –10.5 m OD. The Tilbury III bed forms a persistent unit at –4 to –8.2 m OD, consisting of dark brown monocot peat in a black gyttja matrix. Tilbury IV and V are similar deposits; the former is 30–150 cm thick, and is present here at –2 to –2.5 m OD, whilst the latter locally underlies the modern surface clays. Only three biogenic beds ('peat') interbedded with four 'tidal clay' beds were recorded by Spurrell (1889, fig. 2) and Whitaker (1889) in the adjacent Tilbury Docks excavations. Here, as elsewhere commonly, the Shepperton Gravel is immediately overlain by a green-grey sand bed that varies from 0.5 m to 3 m in thickness.

As already noted, boreholes immediately west of the Dartford Tunnel approach record 9 m of alluvial sediments beneath 2.4 m of made ground (TQ 561758). Similar thicknesses are shown in the Dartford Tunnel southern approach road section (Fig. 46) and were proved by Devoy. Here 0.7–1 m of fine to medium yellow-brown sand rests on the gravel. Three biogenic units occur here, interbedded with the blue-grey silt to clay seen elsewhere. The upper biogenic unit occurs at a mean depth of –0.5 m OD and was correlated by Devoy with his Tilbury IV bed, based on the stratigraphical position and radiocarbon date. The Tilbury III bed is 2–2.5 m thick. The same deposit was also recorded by the author beneath West Thurrock Marshes, on the north side of the river. The lowest biogenic unit at Dartford (Tilbury II), an alder wood peat up to 50 cm thick, occurs at –8.5 m OD.

Only a single 'peat' bed 1.2 m thick is recorded in boreholes upstream beneath Rainham Marshes (TQ 507811), although the alluvial sediments are up to 8.2 m thick here. Likewise, at Dagenham a 2.0 m thick 'peat' unit, interbedded in 4 m of alluvial grey silty clay, is recorded in boreholes west of the Ford Works (TQ 477825: 1 m). The irregular underlying gravel surface and internally complex alluvial sequence are reflected in the dense borehole coverage at Barking where the Roding is confluent with the Thames (Figs. 39, 47). Whitaker (1889) reports sections at the Gallions Sewage Works (c.TQ 4481) and at the Royal Albert Dock, Custom House (TQ 4280) (Fig.38) that exposed 1 m of grey-brown clay resting on up to 1.2 m of peat that in turn rested on either 1.2 m of grey clay or sand. Blandford (in Whitaker, 1889) also observed similar sections at the neighbouring Royal Victoria Docks (TQ 4180).

The surface of the Roding Valley floodplain is also underlain by alluvial sediments equivalent to this unit. The marked decrease in thickness of these sediments upstream in this tributary is interesting, since it parallels that seen upstream in the Thames. This thinning occurs as the sediments pass above 0–3 m OD (i.e. above sea level) at Ilford, since further upstream the sediments are presumably beyond normal tidal influence and are therefore solely of freshwater fluvial origin.

Crossness (Thamesmead: TQ 48158051: 1 m) is Devoy's next locality. This area was previously described by Spurrell (1889, p. 216). Near the sewage works the gravel surface comprises 'a series of ridges and undulating horizontal areas', with 'surface levels of –5 to –6 m OD, cut by channels to –8 to –9 m OD', according to Devoy. This author found only one peat bed, 2.5 m thick (at –1 to –5 m OD), sandwiched between two inorganic clay layers. The peat bed was correlated by Devoy with Tilbury bed III on the basis of height, pollen content and radiocarbon date. In contrast, Spurrell (1889, fig. 3) and Whitaker (1889, fig. 98) found two organic beds, the lowermost resting on 'sand containing freshwater shells and seeds'. The organic beds were rich in tree wood; 'the roots of most of the trees remain in place.' Other plant remains were also abundant and fossil vertebrate and shell material is commonly associated with these beds. Spurrell also notes that the upper peat bed was weathered and included many Roman artefacts and associated bone and shell material. These deposits are present on the south side of the river beneath the Southern Outfall Sewer to Plumstead (Fig. 45) and Erith Marshes (Fig. 46).

The substantial Isle of Dogs meander is underlain by up to 7 m of alluvial sediments comprising the same general sequence (Fig. 41). Two possible biogenic beds occur in boreholes at Cubitts Town (TQ 382787). These could relate altitudinally to Devoy's Tilbury III and IV units. However, older reports (Blandford, 1854; Whitaker, 1889) and modern boreholes at the West India Docks record only a single 'peat' 45–60 cm thick resting on about 1 m of clay that graded downwards into sand, and overlain by mottled brown to grey clay as much as 2 m thick. These deposits thin markedly northwards to only 60 cm beneath the northern end of the docks (TQ 377805). Immediately downstream at Canning Town is the confluence of the River Lea (see below). Boreholes on the eastern bank of the Lea here record 2.98 m of floodplain alluvium resting on Shepperton Gravel (TQ 388810). The deposits comprise a basal brown clay 92 cm thick overlain by 47 cm of 'peat' capped by 1.59 m of brown clay.

These deposits continue under the Surrey Docks at Rotherhithe, but the sequence is much thinner, reaching a maximum of 5 m close to the river (TQ 360802), but thinning rapidly to 3 m 1 km to the south (TQ 361793). Here

Bromehead (1925) reported that the peat contained 'an ancient forest bed with several trees *in situ*'. Towards Southwark the floodplain has been heavily modified by development and in many sections the alluvial sediments are absent, presumably having been removed. In Southwark Evans (1863) recorded 5.8 m of alluvium resting on gravel in excavations at William Street (TQ 320801). Here the basal clay layer was very thin (less than 30 cm) but contained shells and some bones. It was overlain by *c*. 2 m of 'peat' and 2 m of blue clay with freshwater shells. The deposits are well represented beneath Waterloo Station (Gibbard, 1985) and the southern end of Blackfriars Bridge (Fig. 40).

North of the river, the clay–peat–clay sequence continues beneath Branch Road, Limehouse Basin (TQ 362809), at the northern margin of the floodplain. Excavations for the now disused London Docks showed a comparable sequence to that from Southwark with a basal blue clay 30 cm thick, overlain by 1.7 m of 'peat' and capped by silt 2 m thick (*c*. TQ 345805). To the west, the floodplain narrows where the modern river channel abuts higher, older accumulations and therefore no deposits equivalent to this unit are preserved (e.g. Fig. 40). However, thin equivalent alluvial sediments are present in the Fleet valley (Figs. 20, 26) and up to 7.6 m of alluvium has been proved in the Walbrook valley at London Wall (Kennard & Woodward, 1902; Bromehead, 1925).

2.14 Lea Valley floodplain alluvium

As already mentioned, the tributary valleys contain alluvial aggradations whose mode of occurrence is exactly similar to that of the main river valley. The Lea Valley alluvium has rarely been studied regionally, apart from during Geological Survey mapping (e.g. Whitaker, 1889; Bromehead, 1925) and more recently by Hayward (1955, 1956), although some local descriptions have been presented by Wilson (1897), Woodward, (1884) and Holmes (1901, 1902). Throughout the lower Lea Valley, alluvial sediments overlie the Lea Valley Gravel beneath the modern floodplain.

Upstream of the Lea–Thames confluence at Mill Meads, West Ham, a series of boreholes shows the alluvium comprises 30 cm of grey silty clay resting on gravels overlain by 40 cm of 'peat' and capped by 30 cm of mottled grey-brown silty clay (TQ387828: 5.5 m). Similarly, beneath Stratford Marsh 2 m of sediment overlie the gravels. Boreholes at the western end of Stratford High Street also confirm the member is present beneath the floodplain. Detailed records from two boreholes sunk beside the railway at the Blackwall Tunnel northern approach road bridge (TQ 375833: 10 m) suggest that a depression similar to those described by Berry (1979) may occur here. The alluvium is 21.3 m thick, as follows.

	ground surface: 10 m OD
0–5.3 m	Fill.
5.3–6.42	Grey-brown sandy clay and pebbles.
6.42–15.59	Dark grey clay with sand partings.
15.59–16.81	Dark grey silty clay with shells.
16.81–17.12	Grey sand with shells.

17.12–18.96	Mottled brown-grey clay.
18.96–21.78	Dark grey sand and interbedded grey sandy clay.
21.78–25.38	Dark grey silty clay with shells.
25.38–26.60	Dark grey sand.
	gravel and sand

Adjacent boreholes along the northern approach road record normal (i.e. 3.3 m: TQ 376832) rather than these anomalous thicknesses of deposits.

The section along the East Cross Route (Fig. 14) illustrates the irregularity of the gravel upper surface and variable thickness of the alluvial sediment above. The latter appear to be considerably thicker on the western side of the valley in this area. The alluvium is next encountered under Lea Bridge Road where it is again proven by boreholes at TQ 361870 (6 m) to be 2.9 m of brown clay and made ground.

The second cross-section of the valley is along the Victoria Line under Forest Road and westwards to South Tottenham (Fig. 15). As in the East Cross section, the Lea Valley Gravel upper surface is undulating. Again the thickest alluvial sediment seems to be on the western side of the valley. This section at Tottenham Hale is adjacent to a reservoir site (c.TQ 354895) described by Kennard & Woodward (1897) where the highly fossiliferous alluvium was 3 m thick and included peat and shell marl. A Bronze Age canoe was also found in these deposits. Comparable sections have been reported from the Walthamstow Reservoirs immediately to the north, where 1.5 m of shell marl, overlain by 3 m of peat and resting on gravel, were encountered. Woodward (1869) noted that 'the beds above the gravel which forms the floor...vary in thickness and extent over the whole area.'

3
Comparison of the pebble lithological composition of the gravel members

In the descriptions presented in the previous chapter, the pebble lithology of each gravel member and the origins of the material have been discussed. However, if the lithostratigraphical scheme is valid it is necessary to demonstrate as far as possible that the lithological assemblages are sufficiently characteristic to identify the individual members. Such a comparison also provides insight into both local and regional catchment changes that may be present, together with other possible temporal patterns of lithological change.

In order to express and compare multivariate compositional data from the Lower Thames gravel members (see Appendix 2) as simply as possible, the analyses were compared numerically using the multivariate statistical technique, principal components analysis (PCA). Of the techniques available for the comparison of groups of samples consisting of an assemblage of variables, the most useful are those that ordinate all the samples together. Ordination procedures such as PCA seek to represent in a low-dimensional space the similarities between the individual samples. In this technique, samples of similar composition, irrespective of which group they are derived from, will, if the ordination is an accurate representation of the data, be positioned together in the ordination (Blackith & Reyment, 1971; Gordon & Birks, 1974). The geometrical disposition of samples will thus allow the detection both of clusters of similar lithological composition that may originate from the same lithostratigraphical unit and also of groups of dissimilar lithological composition that may originate from different units.

3.1 Principal components analysis

In this analysis, a conventional R-mode PCA of chord distance using a covariance matrix with square root transformation was used to compare samples on the basis of all lithological types present. The results of the analysis are shown in Fig. 51. The positions of individual Lower Thames samples, as represented

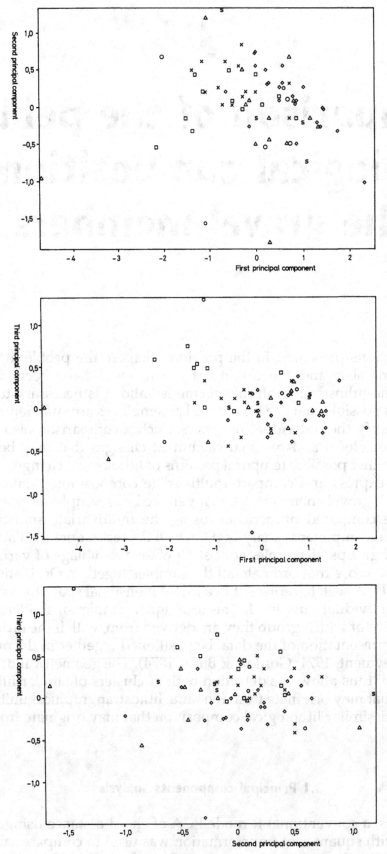

Fig. 51. Principal components analysis of the Thames gravel pebble lithological count data. For explanation see text.

120

by their component scores, are plotted on the first, second and third principal component axes. These remain constant for all Lower Thames members.

In this ordination the first principal component accounts for 62.79% of the total variance. It effectively distinguishes between samples containing significant amounts of angular flint (high positive scores >0.6) and those containing 'Tertiary' or rounded flint (negative scores). The second principal component represents a further 12.60% of the variance and has high positive loadings for quartzite and sandstone (>0.7) and to a lesser extent *Rhaxella* chert (erratics), and negative loadings for Greensand chert, quartz and angular flint. The third principal component represents 7% of the variance and has high positive loadings (>0.9) for non-local erratics (excluding Greensand chert).

The numerical results of the PCA show a generally consistent grouping of samples. They may be summarised as follows:

1. There is a clear separation of those samples containing high quantities of rounded ('Tertiary') flint from those with high amounts of angular flint. The abundance of the rounded flint in the Tertiary bedrock, particularly in the Blackheath Beds in the eastern half of the area (downstream of Purfleet), appears to be reflected in the position of most of the Dartford Heath and many of the Orsett Heath sample sites. This presumably indicates that during deposition of these units the Blackheath Beds were being dissected (as observed at Southfields, Orsett Cock; see section 2.2). A minor trend towards decreasing frequencies of this material is apparent for the Dartford Heath and Orsett Heath samples, but it does not continue in the younger units.

2. The majority of samples from later units (Corbets Tey to West Thurrock Gravel) clearly group very close together indicating, the establishment of a remarkably uniform, stable composition, apart from some very minor local variations. The latter include slightly increased Greensand chert frequencies in samples from areas influenced by south-bank tributaries, such as Swanscombe. They also include samples influenced by incorporation of valley-side bedrock or pre-existing sediment regolith, such as East Court, Gravesend, which is dominated by angular flint of local origin. Apart from these local cross-valley variations, down-valley changes seem to be very small.

3. The high frequency of erratics and greatest sample variability in the Dartford Heath Gravel are emphasised in all three plots. This can also be attributed to the unit being the first Thames' member aligned along the Lower Thames Valley (Chapter 8). Moreover, as the downstream equivalent of the Black Park Gravel (section 2.2), it undoubtedly includes material derived from the Anglian Lowestoft Formation glacial deposits in Hertfordshire, Middlesex, and eastern Essex. The influx of a suite of exotic rock types including *Rhaxella* chert, Carboniferous chert and igneous rocks has been shown by Bridgland (1980, 1986, 1988a) to result from transport of glacial debris into the Thames system during the Anglian Stage. Reworking of glacial material, particularly from the till in the Hornchurch area (section 2.1c), may also explain the higher assemblage variability of the Orsett Heath Member than of the lower units.

3 COMPARISON OF PEBBLE LITHOLOGICAL COMPOSITION

Overall, therefore, the pebble assemblages show a clear pattern in that the earliest assemblages, those of the Dartford Heath Member, have the greatest variability and highest erratic frequency. Both assemblage variability and high erratic content are also found in the Orsett Heath Gravel, but to a lesser extent. However, by the deposition of the Corbets Tey Gravel a stable assemblage had developed and this remains broadly unchanged, although possibly becoming progressively more uniform in the younger members. This trend compares closely with the same trend in the Middle Thames gravels, where all the post-Winter Hill (i.e. Black Park to Shepperton Gravel) members are rich in angular flint (Gibbard, 1985). In the latter area a further trend of decreasing erratic lithologies, accompanied by a proportional increase in angular flint from the older to the younger units, was also identified. In the Lower Thames area, although the older units are richer in erratics, the uniform assemblage seems to have developed earlier and does not appear to show this progressive increase in flint. This slight difference may reflect the proximity of older erratic-bearing gravel spreads and the larger area of Chalk exposure in the Middle Thames compared with the Lower Thames Valley. The development of a generally stable assemblage has also been reported from the downstream equivalents of the Lower Thames Formation members in eastern Essex by Bridgland (1988a).

The discussion above excludes the pebble assemblages from the Epping Forest Formation (high level deposits; section 2.1; Appendix 1). As already mentioned, these gravels comprise restricted assemblages typically containing Lower Greensand chert and rounded 'Tertiary' flint, respectively of Weald and local origin. Small quantities of quartz may also be present and are again thought to be derived from the Weald Lower Cretaceous. These compositions contrast markedly with those of the Lower Thames Formation and therefore separation of the two units is fully supported both by lithological and geological criteria. This separation into two formations closely parallels the situation in East Essex where Bridgland (1988a) independently proposed the same division on the basis of virtually the same criteria.

4
Sedimentary structures and depositional environments

4.1 Gravel units

All the units of waterlain, sorted gravel and sand repeatedly show suites of sedimentary structures that indicate deposition under broadly similar fluviatile conditions throughout the area. It is therefore possible, in general terms, to discuss the sedimentary sequences of deposits together.

Previous investigations of the sedimentary structures of the gravel units have been presented in association with descriptions of fossiliferous fine sediment channel fills in gravels in neighbouring areas. For various reasons these have concluded that the sediments originated in a cold climate river depositional environment. However, detailed facies assemblage studies (e.g. Corner, 1975; Bryant, 1983a, b; Dawson, 1985; Dawson & Bryant, 1987) have not been presented for the Lower Thames region. In the same way as in the Middle Thames area (Gibbard, 1985), a generalised facies description is presented here for the gravel members, based on the sequences examined during this study. The sediments can be assigned to the braided or so-called wandering gravel river environment on the basis of facies models developed by Miall (1977, 1978). Individual gravel members may show minor deviations from the scheme. In addition, the facies descriptions can only be applied to those units that are exposed in large sections where the lateral and vertical relationships can unequivocally be observed.

(a) Descriptive summary

In sections parallel to the valley axis, the sedimentary bodies appear to be elongated in the downstream direction. This phenomenon has frequently been observed from gravel bed rivers (Miall, 1977; Bryant, 1983b) and apparently results from downstream migration of the sediment body. The most commonly occurring facies is massive or horizontally crudely bedded gravel (facies Gm). The gravel clasts show a great size range, but are most frequently of medium to fine gravel (1–30 cm). Larger clasts occur and may reach over 50 cm, but these are rare. The gravels are normally massive, but often show poorly developed horizontal bedding; individual clasts are only rarely imbricated. This may arise because of the relative paucity of blade- and rod-shaped pebbles, the irregular shape of angular flint, and the high sphericity of vein quartz and quartzite rocks (cf. Bluck, 1967). The gravel is generally matrix supported, but clast-supported gravel is common, particularly in the coarser particle sizes. The interstices are filled with silt, sand and granules. Secondary clay may also be present. Individual gravel units are typically up to $c.$ 90–125 cm thick but are frequently superimposed to produce units of greater thickness. Laterally the gravel units may be very persistent and can be traced across quarry faces for over 30 m. They usually rest on an erosional base and are often 'channelled' into the beds beneath.

Broad, shallow channels are often found excavated into the gravel facies. The channels are filled by interbedded tabular cross-bedded gravel and sand (facies Gp and Sp). The channels range from 15 m in width to 1.5 m in depth, but are often much smaller. Individual foresets comprise a complex of pebbly sand, sand and gravel, the grain size showing considerable variation. The bases of such units are normally erosional. The coarsest clasts frequently form a gravel lag at the base. Decreasing upward dip in foreset angle is often observed. Reactivation surfaces (cf. Collinson, 1970) are common. There is a tendency towards upward-fining channel fills, but this trend is seldom completed, since the sequences are often truncated by a reactivation surface and overlain by additional sediments representing the input of fast flowing water.

Trough cross-bedded pebbly sand (facies Gt and St) occurs at all levels in the deposits, normally filling channel-like features cutting across earlier sediments and alternating with gravel facies Gm. The channels are of similar proportions to those mentioned above in cross-section, but individual units, often elongated downstream for 20–30 m, may be as much as 2–3 m in thickness. The base of these deposits is again erosional.

Of the sand facies present only two types are well represented. The most common is the scour-fill sand (facies Ss). This consists of scours, often asymmetric in form, ranging from a few centimetres to about 50 cm and up to about 1.5 m in width. The lenses of sediment often occur within or resting upon bar facies gravel (facies Gm) and comprise pebbly sand. The bedding of the sand is frequently parallel to that of the scour surface. The only other sand facies present appears to be horizontally bedded, often massive sand (facies Sh). These beds are normally of very local extent and up to 20 cm in thickness.

Fine-grained sediments occur very infrequently. Only facies Fl and occasionally Fm are represented by beds of silty clay, clayey silt and organic sediment. The latter is apparently restricted to the lower gravel members. These deposits appear to fill channels or floodplain depressions that have been scoured into underlying sediments or remained as unfilled hollows. They occasionally represent the completion of a fining-upward sediment sequence beginning as facies Sr (ripple-bedded fine to medium sand) and terminating in fine sediment. The fine facies are often grey to brown in colour and contain interbedded sand and even gravel bands. Organic remains, where present, almost invariably include both autochthonous aquatic plant and animal fossils and also allochthonous material washed in during flood events from vegetated stable tracts and neighbouring slopes (Rust, 1972).

Large-scale planar erosion surfaces are a common feature of the gravel members, as noted by Bryant (1983b) in the Upper Thames Valley. They almost invariably underlie medium, to coarse gravel accumulations (facies Gm and Gt) and in any one exposure may be present at various levels. They may be very persistent and are often traceable almost continuously across a quarry face. Where ice wedge casts have been found they often occur beneath such surfaces and indeed may be truncated by them. It is interesting to note that these surfaces are particularly well developed where the contemporary valley was relatively wide: for example in the Ockendon area.

(b) Interpretation

The difficulty of distinguishing between meandering (high sinuosity) and braided (low sinuosity) rivers has been repeatedly stressed by several authors (e.g. Miall, 1977). However, recent models for coarse-grained alluvial sequences (Rust, 1972; Miall, 1977, 1978; Bluck, 1979; Rust & Koster, 1984) have provided a means for discriminating and interpreting the sequences described above.

The main mass of gravels (facies Gm) seems to represent accumulation and migration of low-amplitude longitudinal bars. The gravel was probably laid down in horizontal sheets, but stratification is often obscured by accumulation of finer particles in open interstices. Deposition of such units occurs only at peak discharges, as shown by Hein & Walker (1977). Interbedded sand lenses probably resulted from flow in secondary channels over bar surfaces during low-energy water flow.

Shallow sand and gravel filled channels have been observed in many modern braided rivers (for summary, see Miall, 1977). Minor channels are thought to originate from scour associated with local eddies. Major channels are formed by lateral channel migration at high water stage or bar dissection during falling water conditions (Williams & Rust, 1969; Collinson, 1970; Miall, 1977). These channels, particularly the larger examples, show multiple infills of facies Gt, St and Ss.

Facies St and Sh result from bedform migration. The migration of large-scale ripples or dunes either singly or in groups gives rise to trough cross-bedded sand (facies St; Miall, 1977). Sedimentation by sand waves, of similar proportions to dunes but with straighter crests, produces planar tabular cross-bedding (facies Sp). Such bedforms are produced during upper flow regime flow for which there is ample supporting experimental evidence (Allen, 1970; Friend & Moody-Stuart, 1972). Horizontal bedding (facies Sh) may be formed under lower flow regime in shallow water (Harms et al., 1975) or during the flood stage (Harms & Fahnestock, 1965). In the latter case the bed is formed by streaming of particles across the surface under upper flow regime conditions.

During low-water periods only small-scale bedforms are produced and these are generally restricted to the infilling of minor channels and hollows on bar surfaces by ripple-laminated sand (facies Sr) or in shallow water by horizontally bedded sand (facies Sh). Sand wedges may develop during falling water stages at bar margins. Such wedges normally show planar cross-bedding. Silt, clay and organic material (facies Fl or Fm) accumulate in abandoned or partially abandoned flow channels in standing or trickling water, especially in topographically higher areas of the braidplain (Williams & Rust, 1969; Rust 1972). In spite of their apparently low preservation potential (cf. Cant, 1976), these beds are occasionally found.

Braided rivers commonly only occupy part of the valley bottom at any particular time and large areas of the braidplain may become temporarily stable and may even be colonised by vegetation (Williams & Rust, 1969; Rust, 1972; Miall, 1977; Rust & Koster, 1984). During exposure these surfaces are subjected to subaerial weathering, and ice wedge polygons and cryoturbation may develop in favourable localities, particularly on higher stable tracts (Washburn, 1968; Bryant, 1983b). Renewed channel migration will degrade such a surface, terminate subaerial weathering and result in talik development leading to degradation of ground ice phenomena. Infill and truncation of ice wedge casts developed in the surface, together with the truncation and burial of valley-side mass flow deposits and abandoned channel fills may therefore be expected at this time. The extensive planar erosional surfaces have been attributed by Bryant (1983b) to truncation of fluvially inactive areas, raised slightly above the active areas of the river by large-scale channel migration. The association of truncated infilled ice wedge casts and other periglacial phenomena, channel fills and earlier sediments beneath such surfaces supports this interpretation.

The assemblage of sedimentary facies and structures present in the Thames gravel members broadly comprises a sequence of interposed channel fills and superimposed longitudinal bars. This assemblage most closely approximates to the Donjek depositional model proposed by Miall (1977, 1978) from the middle reaches of the river (Area 2 of Rust, 1978a; facies model G_{III} of Rust, 1978b). The facies observed in the present study closely resemble those described by Bryant (1983b) from the Upper Thames.

Taken together all these factors indicate that the deposits accumulated in a braided river environment. The consistent association with climatic indicators such as ice wedge casts, solifluction deposits and cryoturbation structures suggest aggradation under a predominantly periglacial regime. This is strongly supported by the facies present which have been shown to typify nival-flood dominated rivers in the modern Arctic (Bryant, 1983a).

4.2 Mass flow deposits

Near contemporary valley sides, lenticular wedge or tongue-shaped pebbly clay or 'rubble chalk' diamictons are often found, both interstratified and overlying fluvial sediments. These diamictons vary considerably in thickness from a few centimetres to over 2 m and have erosional bases often showing incorporation of clasts or sediment from beneath. The overlying fluvial sediments always show marked erosional contact with these diamictons. The pebbly clay beds are usually massive, but the sediment is often poorly mixed and silt laminae or sand stringers may occur. The latter rise down the dip of the deposit, may be distorted around pebbles, and may show flow-type banding parallel to the lower bounding surface. Downslope clast orientation is predominant (cf. Watson, 1969).

The 'rubble chalk' or 'coombe rock' (Reid, 1877) beds comprise angular chalk fragments and flints in a putty-like matrix of mechanically fragmented chalk. In the unmodified state this material may be very difficult to distinguish from the upper few metres of blocky bedrock chalk *in situ*. However, the material shows considerable variability and may in extreme cases, following from incorporation of underlying material and decalcification, be difficult to separate from poorly sorted, massive fluvial gravel. This material is well known in the Lower Thames valley where it cloaks bedrock chalk slopes. The deposition of this material was even referred to a specific time period by King & Oakley (1936), but this is clearly an oversimplification, to judge from its relationship to the fluvial sequences described in Chapter 2.

These pebbly clay or 'rubble chalk' diamictons were most probably deposited by mass flow or solifluction rather than fluvial processes. Their markedly local clast content, together with the abundance of frost-shattered pebbles, indicates a local source subject to cold climate weathering. The mixture of these pebbles in a non-durable, unsorted matrix strongly suggests intermixing of material from the immediate vicinity. Weathering of bedrock under a frost-dominated climate would provide unstable slopes, and downslope flow would be expected (cf. Galloway, 1961). For further discussion of solifluction deposits in south-east England see Hutchinson (1991).

4.3 Sand and fine sediment association

The sites described in section 2.7 include sedimentary sequences that are markedly different from those discussed above. Several of these sequences (section 2.7b, c, d, e, f, g, i and l) predominantly comprise fine sediments, i.e. silts and clays, either bedded or massive, and occasionally laminated and often associated with current bedded sand. These fine sediment associations are generally refered to as 'brickearth'; however, they differ in origin from the clayey silt sediments with the aeolian component (section 2.10) in that they are well stratified and therefore waterlain. They also commonly include fossil assemblages, both plants and animals. Considerable attention has been paid to these sediments, and in a recent review Hollin (1977) recognised three varieties: 'a laminated, relatively sandy sediment at Crayford, Little Thurrock (Grays Thurrock) and Purfleet, probably of intertidal origin', 'a massive clayey brickearth at Crayford, Little Thurrock, Aveley and West Thurrock' and the colluvial material often including an aeolian component (discussed in section 2.10).

In this section some thoughts on the depositional setting of this waterlain fine sediment association are presented in an attempt to provide a potential basin-wide, rather than single-site, interpretation. This offers a firmer foundation for the interpretation of the contained fossil assemblages than those previously available. This description does not include all the sites described in section 2.7, since the depositional environments of several are thought to be predominantly fluviatile (i.e. section 2.7a, i, j, k).

(a) Descriptive summary

In sections the deposits appear to show no preferred orientation of the sedimentary bodies. The most commonly recorded facies is interbedded or interlaminated clays or silty clays and sands. The clays or silts may be grey, brown or green in colour, are <1 to 10 cm in thickness, are frequently finely laminated, have gradation bases with the underlying sands and are occasionally disturbed by loading or water release structures. The sands are >20 cm thick, frequently current bedded (either tabular or ripple cross-laminated) and have erosional bases. They occasionally include fine pebbly horizons (including clay 'rip-up' type clasts), abundant mollusc shells and other fossil animal remains.

A particular variety of the laminated sediments is finely laminated sandy silt alternating with silty clay up to 3–5 mm in thickness, present at Aveley and Purfleet, for example (section 2.7d and f). As noted above, these horizontally laminated units contain much small-scale bedding structure, including water-release structures and microchannels, the latter filled with sand, granules, shell fragments, etc.

Massive clay or silt is the next most common facies. This sediment varies in thickness from 1 cm to 3 m and is generally light grey to brown in colour. Its base is conformable on the underlying sediments and it frequently contains scattered shells.

Of similar frequency to the massive units are beds of silty sand to sand, either finely horizontally stratified or laminated and rarely ripple bedded. These sands do not appear to fill channels, but resemble sheet-like accumulations 1–2 cm in thickness. The base of each of the sheets is either conformable or slightly erosional. They are often built up into thick beds that are continuous laterally, but in places appear to interdigitate with the massive fine or laminated sand–silt sediments. Current structures are rare in these sand units, but where present are usually tabular cross beds.

At the base of these sediment sequences, where they do not directly overlie older gravel and sand member sediments, a pebble lag occurs (e.g. West Thurrock; section 2.7c). This lag is usually only a few centimetres in thickness and consists of poorly rounded flint clasts. Artefact material and bone may occur in this lag.

(b) Interpretation

These waterlain, fossiliferous fine sediments clearly indicate deposition generally under slow velocity water flow. Such conditions may exist in the colluvial, fluvial, estuarine and marine environments. As already discussed, colluvial sediments are abundant in the area and they do not resemble those described here. Fluvial processes certainly controlled sedimentation at several localities described in section 2.7, but most of the sequences appear very thick (some over 10 m) and do not resemble those known from the present floodplain upstream of the estuarine area. Moreover, they clearly lack the facies assemblages typically found in river systems of braided, wandering or meandering forms (cf. Walker & Cant, 1984, and above). This leaves the estuarine and marine environments. As already mentioned, Hollin (1977) concludes that the laminated fine sediments are probably of intertidal origin. Could the remaining sequence also be related to tidal or shallow marine sedimentation? This question is particularly relevant since the area lies within the modern Thames Estuary and therefore might reasonably also be expected to have been so in the recent geological past.

The detailed investigation of the estuarine and shallow marine environments has produced a literature too large to review in detail here. However, fortunately an excellent series of reviews have been published by Davis (1985) and it is to this volume that primary reference is made for sequence comparisons. In view of all the evidence presented above, the only environment in which all the fine sediment association lithofacies described can have accumulated is the estuary (cf. Nichols & Biggs, 1985). The description below is modified from the review by these last mentioned authors.

The lithofacies of estuarine sedimentation generally reflect submergence, transgression and potentially later regression in a valley whose form is controlled by that of the pre-existing river valley system. To understand the pattern of sedimentation it is important to appreciate that laterally equivalent contemporary sediments of varying facies may come to be seen in vertical succession as sedimentation proceeds (so-called 'Walther's Law': Middleton, 1973).

According to Nichols & Biggs (1985), the axial lithofacies sequence upstream to downstream in an estuary comprises: (a) estuarine fluvial, (b) estuarine, and (c) estuarine marine. Therefore, transgression will theoretically result in the deposition of lithofacies (a), then (b) followed by (c) at any one locality. Clearly, lateral as well as vertical changes will occur such that mid-channel sediments pass into subtidal flats and finally shoreline deposits. Sediment lithofacies are controlled both by short-term (e.g. tides) as well as long-term (e.g. sea level) dynamic processes. If the sedimentation rate exceeds local sea-level change, then the estuary will fill with sediment and the fluvial deposits may prograde over the estuarine sequence (cf. Roy, Thom & Wright, 1980). In this situation lithofacies (a) may come to overlie (b) and so on.

The potential complexity generated by the three-dimensional evolution of an estuary would therefore be difficult to unravel if sediment sequences were well preserved. However, in the Lower Thames area the sequences under investigation are poorly preserved, mostly as eroded remnants perched along the valley side. Moreover, many of the localities are no longer available for study. Consequently, little more than a general interpretation can be attempted here. Nevertheless the importance of the deposits to the sedimentary history of the valley requires that they be interpreted as fully as possible.

In the fluvially dominated parts of estuaries the volume of sediment transported and the dispersal pattern is controlled by river flow. Suspended sediment is diluted downstream and accumulates partly in marshes, resulting in massive clay and silt often rich in organic material. Suspended sediment is also carried downstream and deposited in the less active areas of the channel and on bars or is transported seaward. Sand accumulates on channel floors during periods of higher river discharge, whereas fines are laid down during low discharge periods. Such sequences may occur as part of the normal tidal cycle. This results in the accumulation of the commonly observed interbedded sand and fine sediment facies. Individual sand deposits a few centimetres in thickness may occur in massive fine sediments as a result of current reworking. Longitudinal bars are fed by sand bed load, but often include clay lenses or pebbles.

Dispersal of fine sediment into the main estuary occurs by water circulation, controlled both by the river and tidal effects. This gives rise to the accumulation of thicknesses of often very mobile fines. Channel infill facies from this area may be laminated or wavy-bedded clays with occasional sand lenses or bands. Lower estuary sediments are similar except that they are frequently heavily bioturbated and therefore appear as massive shell-rich silts or clays. In contrast, shallow water margins are areas of erosion by waves and therefore often occur as rippled sand with clay pebbles and erosional contacts, while in protected areas laminated clays and silts are deposited.

Sedimentary structures in the fluvial–estuarine (tidal-controlled) area are characterised by ripple-bedded sand that migrates both upstream and downstream, giving rise to the classic 'herring-bone' cross-bedding (Reineck & Singh, 1980). This bimodal dispersal is controlled by the channel morphology.

Shell and plant material is common throughout estuarine sequences, either distributed throughout the sediments or as discrete bed concentrations. Plant remains are generally concentrated in marginal or protected shallows or fringing marshes, whereas Mollusca may inhabit large bottom areas. Short-term effects such as tidal surges can cause catastrophic death of freshwater Mollusca and other bottom-dwelling organisms and can give rise to beds with high shell concentrations. Equally, flood events can sweep large quantities of terrestrially or shallow-water derived plant and animal remains into relatively deep water settings (cf. Nichols & Biggs, 1985).

Lastly, the basal gravel lag may reflect local reworking of slope and superficial regolith material that was redistributed by waves during transgression and tidal activity.

In conclusion, therefore, it seems that all the facies described from the fine sediment association can be accommodated in the lithofacies assemblage of an estuary that presumably formed during a period or periods of submergence of the valley by marine transgression. The interpretation of the individual site sequences will be presented below. However, the lithofacies assemblage suggests deposition in the fluval estuarine and estuarine zones, since no sediments of the estuarine marine facies are present.

5
Vertebrate faunal assemblages

For over 150 years vertebrate remains have been recovered and collected from localities in the Lower Thames Valley. Many of these finds have been preserved in public or private collections, whilst others have long since been lost. Collections are normally dominated by large vertebrate material which was easily recognised during excavation, smaller remains having been often overlooked. By far the largest numbers of finds were made during the nineteenth century when digging was undertaken by hand. Since the advent of machine working, a severe decline in the frequency of finds has taken place.

Although the Lower Thames Valley contains some of the most productive vertebrate localities in the country, a complete synthesis has never been published. The most detailed study is that of Stuart (1976), but this concentrated on only five sites for which dating was relatively well established at the time. Stuart concluded that these five sequences each represented a part of a single temperate (interglacial) event (the Ipswichian Stage) and that changes in the faunal assemblage corresponded to similar changes recorded by the palaeobotanical evidence (Chapter 6). In contrast, Sutcliffe (1975, 1976, 1985) has repeatedly suggested that the climatic sequence of the Pleistocene was more complex than generally thought on the basis of the comparison of vertebrate assemblages from sites in the area.

The discussion concerning the dating of individual sequences and their contained fauna continues to the present day (cf. Bowen *et al.*, 1986). However, the key issue bearing on the interpretation of these fossil assemblages is their stratigraphical position, since until now the stratigraphy of the Lower Thames has been vague and much oversimplified. The interpretation of these vertebrate assemblages has therefore been fraught with pitfalls. The purpose of this section is to attempt to clarify the stratigraphical position of the assemblages and to offer possible explanations of their potential significance for the palaeoenvironmental reconstruction of the area.

The fossil assemblages are summarised in Table 1, which includes finds from both the gravel members as well as the important interglacial and associated units.

5 VERTEBRATE FAUNAL ASSEMBLAGES

Table 1. *Vertebrate faunas from the Lower Thames sediments*

Unit or site	Swanscombe Lower Gravel	Swanscombe Lower Loam	Swanscombe Middle Gravel	Ingress Vale	Orsett Heath Gravel	Corbets Tey Gravel	Mucking Gravel	Northfleet
Homo sp.			+					
Oryctolagus curriculus	+	+						
Legus sp.	+	+	+					
Crocidura suaveolens								
Arvicola cantiana	+			+				
Microtus sp.								
Microtus agrestis			+	+				
Microtus arvalis								
Microtus gregalis	+							
Microtus œconomus	+							
Clethrionomys glareolus			+	+				
Pitymys arvaloides	+							
Lemnus				+				
Sorex araneus				+				
Canis lupus	+	+	+	+				
Spermophilus undulatus								
Ursus spelaeus	+							
Ursus arctos								
Martes cf. *martes*	+							
Crocuta crocuta								
Panthera leo	+	+	+					
Felis sylvestris	+							
Palaeoloxodon antiquus	+	+	+	+				
Mammuthus primigenius						+	+	+
Equus ferus	+	+	+	+	+			+
Coelodonta antiquitatis								
Dicerorhinus hemitoechus	+			+				+
Dicerorhinus kirchbergensis	+	+	+	+				
Sus scrofa	+			+				
Trogontherium cuvieri				+				
Castor fiber	+							
Hippopotamus amphibius								
Cervus elaphus	+	+	+	+				
Capreolus capreolus	+							
Dama dama	+	+	+	+				
Megaceros giganteus	+	+	+					
Rangifer tarandus								
Bos/Bison								
Bos priscus						+		
Bos primigenius								
Ovibos moschatus								
Saiga tartarica								
Apodemus sylvaticus				+				
Delphinus sp.				+				
Dicrostonyx torquatus								
Lemmus lemmus								
Talpa cf. *minor*	+							
Macaca sp.	+							
Emys orbicularis				+				

132

5 VERTEBRATE FAUNAL ASSEMBLAGES

Grays Thurrock	West Thurrock	Aveley	Crayford–Erith Gravel	Crayford–Erith "Lower brickearth"	Crayford–Erith "Upper brickearth"	Ilford Uphall	Ilford High Road	Shacklewell Lane	East Tilbury Marshes Gravel
+						+			
+						+	+		
+	+				+	+	+		
+	+	+			+	+	+		
+	+	+	+	+	+	+	+		?+
+	+	+	+	+	+	+	+		+
+	+	+	+	+	+	+	+		+
+	+	+	+	+	+	+	+		
+	+	+	+	+	+	+	+		
+	+	+	+	+	+	+	+		
+	+	+	+	+	?	+	+		+
+	+	+	+	+	+	+	+		
+	+	+	+	+	+	+	+		+
+	+	+	+	+	+	+			
				+					
+	+	+	+	+	+	+			
+	+	+			+	+			
				+					
				+					
+									

133

5 VERTEBRATE FAUNAL ASSEMBLAGES

5.1 Taphonomy and preservation

The taphonomy of vertebrate remains in river sediments has been considered by Hanson (1980) and Stuart (1982) to whom reference should be made for detailed discussion. Both, however, stress the need to consider the various stages through which skeletal parts pass before they become fossilised in sediments. In this area partial or complete skeletons are very rare, but have been found in fine sediments. The most well-known discovery was of two elephant skeletons lying virtually one above the other at Sandy Lane, Aveley. Here the lower straight-tusked elephant *Palaeoloxodon antiquus* lay in sediments of Ipswichian Ip IIb age, whereas the upper mammoth *Mammuthus primigenius* was of IpIII age (West, 1969; Stuart, 1976). An almost complete elephant skeleton was also recovered from Grays Thurrock (Morris, 1836). Partially articulated remains of an aurochs *Bos* or extinct bison *Bison* were also reported by Prestwich (1855) from Shacklewell Lane, and part of a *M. primigenius* skeleton was found in the Ilford High Road ('Cauliflour') pit in 1897–8 (Hinton, 1900). The discovery of complete skeletons suggests that the animals died in or close to the depositional environment, by becoming stuck in muddy sediment or drowned, for example (cf. Stuart, 1976, 1982).

However, as Stuart (1982, p. 65) notes, disarticulated remains are much more common. These have been transported to the depositional site by the river and/or periglacial slope processes. Animals are attracted to rivers to drink and feed on riverside vegetation. The corpses of animals that died on the river floodplain could have been attacked by scavengers and their skeletons left in a disaggregated form. Floating cadavers could also have been carried by the stream until they became waterlogged and sank, disaggregated by continual buffeting or until they came to rest on an obstacle such as a point bar or gravel braid. Winter freezing under cold climates may have tended to preserve corpses so that skeletal parts may remain unaltered for many years. Slope processes such as cold climate solifluction and slope wash of regolith, in treeless environments, could potentially carry significant material. This process, combined with bank erosion, could therefore provide a further important contribution to river sediments.

The problem of reworking of previously deposited vertebrate remains must be addressed. Erosion of the river bed and banks during downcutting events, including channel migration and slope denudation, may result in the incorporation of previously fossilised material from the underlying bedrock, as well as pre-existing Pleistocene sediments. There is no doubt that this occurs, but it may only be detected when the remains of temporally or ecologically incompatible creatures or remains in very different states of preservation are found in the same stratigraphical unit.

Nevertheless, once bones, teeth or antlers enter the river they become sedimentary clasts subject to the same processes as all other sedimentary particles in the fluvial system. Thus most of the vertebrate material found in coarse river sediment is relatively large and this selection must result from sedimentary sorting. As Gibbard (1985) discusses, large vertebrate material has been found from almost all facies of braided river sediments, but particularly gravel braid (facies Gm) and channel scour and fill deposits (facies Gt). In meandering rivers, vertebrate material may occur as concentrations or single particles in flood sand units, basal 'lags', point bar sediments and fine sediment channel or floodplain depression fills. Articulated material is most likely to occur in subenvironments in which low velocity flow predominates, i.e. the areas of fine sediment accumulation. In estuaries, as elsewhere, remains will be transported like other sedimentary particles.

5.2 Faunal assemblages

The faunal list in Table 1 was compiled from various sources, most of which are mentioned in the description of members in Chapter 2. The remaining records are from Stuart (1982), Carreck (1976), Sutcliffe (1964) and Zeuner (1959). The author is not responsible for the vertebrate fossil identifications which are based on those used in the original publications or lists. However, the nomenclature has been modified to present-day classification following Stuart (1982). In addition, the identification of *Bison priscus* and *Bos primigenius* is confused, older records referring simply to 'ox'. The separation of the remains of these species is very difficult and therefore where any doubt was possible the records are assigned to *Bos/Bison*.

(a) Cold environment assemblages

As noted in the Middle Thames region (Gibbard, 1985, p. 104), the number and diversity of preserved remains appears to be generally proportional to age: i.e. more remains are recorded from younger deposits. This probably results from increasing carbonate solution with increasing antiquity of the deposits. It undoubtedly explains the absence of vertebrate material from older, higher units.

However, in contrast to the Middle Thames where all the main gravel member sediments from Boyn Hill to Shepperton Gravel have yielded some vertebrate material (Gibbard, 1985), finds from the Lower Thames gravel members are extremely rare. To the author's knowledge no vertebrate remains have been collected from either the Dartford Heath or the Orsett Heath gravels. By contrast, some faunal material has been collected from the Corbets Tey/Lynch Hill Gravel. It includes what Wymer (1968, p. 289) describes as 'either a complete skeleton, or more likely, some molars or tusks' of an elephant from the pit on King's Cross Road from where 'the first palaeolith recorded in Britain' was also recovered. Two workings near Wanstead have also produced horse *Equus* and *Bos/Bison*, according to Hinton (1900).

The only records of vertebrate material that can be definitely related to the Mucking Gravel are the finds from the gravel underlying the Crayford 'brickearth' sediments. These certainly included elephant teeth and rhinocerid remains, according to Spurrell (1885) and Kennard (1944) (see also below). Possibly equivalent deposits overlying the 'rubble chalk' at Northfleet (section 2.7a) have also yielded faunal remains, including mammoth, horse and woolly rhinoceros *Coelodonta antiquitatis*.

The most diverse fauna from the gravel members is that collected from the East Tilbury Marshes Gravel at Greenwich by Boulger (1876). Here *Equus*, *Bos/Bison*, rhinoceros (?*Coelodonta antiquitatis*) and 'two or three species of *Cervus*' were found. Importantly, this assemblage also included *Hippopotamus*. The same unit has also yielded 'fragments of elephant's tooth, and a whole tooth of rhinoceros' from Deptford (Pattison, 1863). No vertebrates are recorded from the West Thurrock or Shepperton gravels in the area.

The Lea Valley has produced faunal material, particularly in association with the 'Arctic Bed'. The remains include the small vertebrates arctic lemming *Dicrostonyx torquatus*, Norway lemming *Lemmus lemmus*, northern vole *Microtus oeconomus* and tundra vole *M. gregalis*, frog *Rana* sp., toad *Bufo* sp. and lizard *Lacerta vivipara*. The large mammals include *M. primigenius, C. antiquitatis, E. ferus, Bos/Bison* and reindeer *Rangifer tarandus* (Warren, 1912; Stuart, 1982). This late Middle to Late Devensian fauna is typical of those from the period and appears to be completely consistent with the palaeobotanical evidence for a herb-dominated park or grass sedge tundra flora (Allison *et al.*, 1952; Stuart, 1982).

The gravel members therefore contain generally similar assemblages. All the main members have produced elephant material, either certainly or probably referable to *M. primigenius. Equus ferus* is also common. This compares closely with the Middle Thames area (cf. Gibbard, 1985) where horse was present in all but one of the major members. Indeed, the assemblages from the gravel members all correspond very closely with the records from their Middle Thames equivalents. Likewise, the faunas support the sedimentary evidence for gravel and sand aggradation under cold climates. They suggest an herbaceous steppe or tundra-like vegetation on the surrounding areas. Similar assemblages are known from some 'brickearth' localities where the sediments are thought to be of periglacial colluvial origin (see below).

Biostratigraphically, the faunas are of little value, with the exception of the find of hippopotamus in the East Tilbury Marshes Gravel. *Hippopotamus* sp. remains are known only from the late Cromerian and the Ipswichian stages, but not from cold stages or the Hoxnian in Britain (Stuart, 1984). The complete lack of Cromerian age deposits in the Middle and Lower Thames makes it reasonable to assume that all the hippopotamus remains were probably Ipswichian. The occurrence of this 'interglacial' animal in cold stage sediments indicates that the remains must have been reworked from local interglacial deposits where it is well represented (see below). The gravel aggradations containing these remains must therefore post-date the Ipswichian Stage and therefore must be Devensian (cf. Gibbard, 1985).

(b) Temperate environment assemblages

In contrast to the paucity of vertebrate material from the gravel members, the finer sediment sequences in the area have been amongst the most productive open sites in the country. The assemblages from these sites are also shown in Table 1.

At Swanscombe (Sutcliffe, 1964; Stuart, 1982) the sequence is divided into four subunits (section 2.3), the lower three having yielded numbers of vertebrate finds. The Lower Gravel is itself divisible into lower and upper subunits. As noted above, Conway (1973) recorded fallow deer *Dama dama, Equus ferus* and *Bos primigenius* from the basal gravel. This assemblage indicates that mild conditions were already established at this time. The upper part contained a more diverse assemblage including macaque monkey *Macaca* sp., beaver *Castor fiber*, a range of voles, marten *Martes martes*, wild cat *Felis sylvestris*,

5.2 FAUNAL ASSEMBLAGES

Palaeoloxodon antiquus, extinct rhinoceroses *Dicerorhinus kirchbergensis* and *D. hemitoechus*, various deer and wild boar *Sus scrofa*. An almost identical fauna is also found in the Lower Loam. Stuart (1982, p. 122) comments that this fauna 'is consistent with regional mixed oak forest', but includes some 'taxa consistent with probably rather local, more open habitats'. Fish and bird material is also known from these units.

The finds from the Middle Gravel have not generally been separated into those from the lower and upper parts of this unit. In general the fauna from this bed contains many of the elements present in the beds beneath; however, the finds are fewer in number (possibly reflecting the poor preservation of material higher in the sequence). 'Most of the indicators of more open vegetational conditions, e.g. horse, giant deer...continue from the Lower Gravel and Lower Loam, joined by mountain hare *Lepus timidus* and...Norway lemming *Lemmus lemmus*' (Stuart, 1982): together they suggest a deteriorating climate. The continued occurrence of forest elements such as *P. antiquus* and *D. dama* suggested to Stuart (1982) that woodland was still present (contrary to the molluscan evidence of Kerney, 1971). However, Stuart does not consider the possibility that these remains could have been incorporated by reworking of the immediately underlying or possibly adjacent sediments. This highly probable occurrence would not only explain the vertebrate finds, but also sparse molluscan evidence for contemporary forest.

The Upper Middle Gravel also contained fragments of a human skull, whilst the higher, 'Upper Gravel' yielded remains of the arctic musk ox *Ovibos* sp. (Waechter & Conway, 1977).

The flow channel shelly sands at Ingress Vale, probably contemporary with the Lower Loam floodplain development at Barnfield Pit, have also yielded vertebrate remains. In general the assemblage closely corresponds to that from the Lower Loam, but includes extinct beaver *Trogontherium cuveri*. This species is known from several Hoxnian Stage sites. Additionally, a vertebra of a dolphin *Tursiops* sp., possibly from an animal beached in the channel, has been collected (Sutcliffe, 1964; Stuart, 1982).

Overall, the vertebrate fauna from the Swanscombe Member complements the evidence from the sediments and molluscan fauna for environmental change from fully temperate in the lower part into cold climates in the upper part of the sequence.

The faunas of the complex of fine sediment units, predominantly interglacial sequences referred to the Ipswchian Stage by Stuart (1976) and related sites (section 2.7), will now be discussed. Detailed arguments for the age of individual sequences, or parts of sequences, will be presented after the palynological evidence (Chapter 6), but disagreements relating to the vertebrate fauna will be mentioned here.

The sections in the Ilford area, described in section 2.7i, appear to comprise a complex of sequences rather than a simple single sedimentary unit. The vertebrate fossil discoveries were almost exclusively from the Ilford Sands and Silts rather than the overlying Langley Silt Complex 'brickearth'. At the Uphall site a vast collection of material, mostly from the so-called '*Corbicula* bed', included a skull of *Dicerorhinos hemitoechus* and the overlying gravel and sand yielded a *Mammuthus primigenius* skull (Dawkins, 1867) together with a series of other remains. According to this author, 'None of [the faunal remains]..presents any traces of rolling.' Hinton (1900) also notes that *Hippopotamus amphibius* was found here. At the High Road pit a very similar fauna occurred. By comparison with the Seven Kings sequence (West *et al.*, 1964), Stuart (1976) concluded that this temperate fauna dated from the second half of the Ipswichian Stage (Ip III to IV) and indicates the local occurrence of forest, but with extensive areas of herb-dominated vegetation.

There has been considerable discussion (much unpublished) regarding the age of the Ilford Sands and Silts, based on the view of some vertebrate experts that some of the mammoth remains show 'primitive' characteristics and may therefore indicate an older age than Stuart suggests (e.g. Sutcliffe, 1976, 1985). Carreck (1976), however, considers that the Ilford material resembles the 'true mammoth' from West Thurrock and Crayford. Such arguments are beyond the scope of this work, although it is possible that some of the material at Ilford may have been incorporated by reworking from higher units preserved north of the site (e.g. Corbets Tey Gravel).

The faunal remains from Sandy Lane, Aveley, have already been mentioned. However, those from the extensive workings on the opposite side of the river at Erith to Crayford have again yielded a large range of finds. The 'lower brickearth' unit contained most of the fossil finds (section 2.7e) and, according to Chandler (1914, p. 69), most of the Mammalia were collected by Spurrell from Stoneham's Pit. The faunal assemblages from Crayford have been considered by Stuart (1982) and his comments form the basis of those presented here.

The 'lower brickearth' fauna is uniform throughout the deposit, the finds occurring up to and including the so-called '*Corbicula* bed'. It resembles that from Ilford in that it includes temperate types *Bos primigenius* and *Dicerorhinus hemitoechus*. However, the assemblage is characterised by the occurrence of some so-called cold stage species, such as Norway lemming *Lemmus lemmus*, arctic lemming *Dicrostonyx torquatus*, northern vole *Microtus oeconomus*, woolly rhinoceros *C. antiquitatis* and musk ox *Ovibos moschatus*. These 'cold habitat' mammals contrast with the Mollusca which are notably of temperate and southern affinities. In comparison with other last interglacial sites in the Thames Valley, the Crayford fauna appears somewhat transitional between those of the second half of the Ipswichian and those of the Devensian. For this reason, Stuart (1982) considered the assemblage and therefore the deposits to be of late Ipswichian (Ip IV) or earliest Devensian age.

The overlying solifluicted 'upper brickearth' is thought to be certainly cold stage (Devensian) since it contains the typical 'cold faunal elements' *M. primigenius*, *C. antiquitatis*, *E. ferus* and deer.

Evidence that the basal gravel (beneath the 'lower brickearth') may not have been entirely cold stage, but in part included a channel fill or 'lag' of temperate origin, is suggested by the occurrence of *P. antiquus* and possibly hippopotamus, according to Carreck (1976).

The vertebrate fossils from West Thurrock have been reviewed by Carreck (1976). The main collection from the northernmost sections here was by Abbott (1890), Hollin's site WT3 (1977; section 2.7c), from the thick current-bedded sands. The mammals found include *M. primigenius*, *C. antiquitatis*, *D. hemitoechus*, *D. kirchbergensis*, *Hippopotamus amphibius*, red deer *Cervus elaphus*, *E. ferus*, giant deer *Megaceros giganteus* and *Bos/Bison*. As Carreck (1976) points out, this assemblage is clearly of temperate type and suggests the occurrence of riverside grassland with temperate forests beyond. The fauna resembles those from Crayford and Ilford, but on the basis of the presence of straight-tusked elephant and hippopotamus, he places the deposits slightly earlier than the Crayford 'lower brickearth' in the Ipswichian, probably Ip IV. However, if Stuart (1976) is correct that hippopotamus is not present in the Ipswichian after Ip III, a slightly earlier date would seem more appropriate.

Grays Thurrock has also yielded a similar, although richer, fauna including *Palaeoloxodon antiquus, Mammuthus primigenius, Hippopotamus amphibius, Ursus, Hyaena, Macacus, Dicerorhinus hemitoechus, Coelodonta antiquitatis, Megaceros, Microtus agrestis*, etc. Once again there can be little doubt that this fauna must also be of middle to later Ipswichian age by comparison with those above.

Overall, therefore, there appears to be considerable evidence for similar, late Ipswichian (Ip III–IV) faunas from several sites in the Lower Thames Valley. Earlier Ipswichian assemblages are rarer, the only sites being Aveley, the possible Crayford gravel, the Shacklewell Lane finds and Trafalgar Square (Stuart, 1982; Gibbard, 1985).

6
Palaeobotany and biostratigraphy

Palaeobotanical analyses have been undertaken to provide both a biostratigraphical framework for the inorganic deposits and a basis for palaeoecological interpretation of the fossiliferous sediments. The studies of several of the sites have been undertaken with co-workers and the full descriptions of the palaeoecology will be published separately as appropriate. A synopsis of the palaeobotany of sites in the Lower Thames region is presented here. The localities are discussed in geographical order from east to west.

Pollen preparation followed standard chemical methods used in the Subdepartment of Quaternary Research, University of Cambridge (West, 1968), but included the use of sodium pyrophosphate (Bates, Coxon & Gibbard, 1978). The pollen sum in all cases is total land pollen and spores excluding aquatic taxa. Pollen conventions follow Andrew (1970) together with some types listed in Birks (1973).

6.1 Grays Thurrock

Detailed palynological investigations have been carried out from the borehole (Grays A) described in section 2.7b. The pollen diagram is shown in Fig. 52. The diagram cannot be subdivided and therefore comprises a single pollen assemblage biozone (p.a.b.) throughout. No pollen was preserved in samples above 6.20 m.

6.20–10.75 m *Quercus–Alnus–Pinus* p.a.b.

Throughout this zone the spectra are dominated by temperate deciduous trees including *Quercus*, *Alnus*, *Corylus* and to a lesser extent *Acer*, *Betula*, *Fraxinus*, *Ulmus* and *Tilia*. The prevalence of fully temperate conditions is confirmed by the pollen of frost-susceptible *Hedera* and *Ilex*. Pollen of the conifers *Pinus* and to a lesser extent *Picea* are consistently present. An important component of the spectra is grass and herb pollen, indicating that grassland communities were common locally on damp and dry calcareous ground. Occasional records of the pollen of plants associated with disturbed or trampled ground,

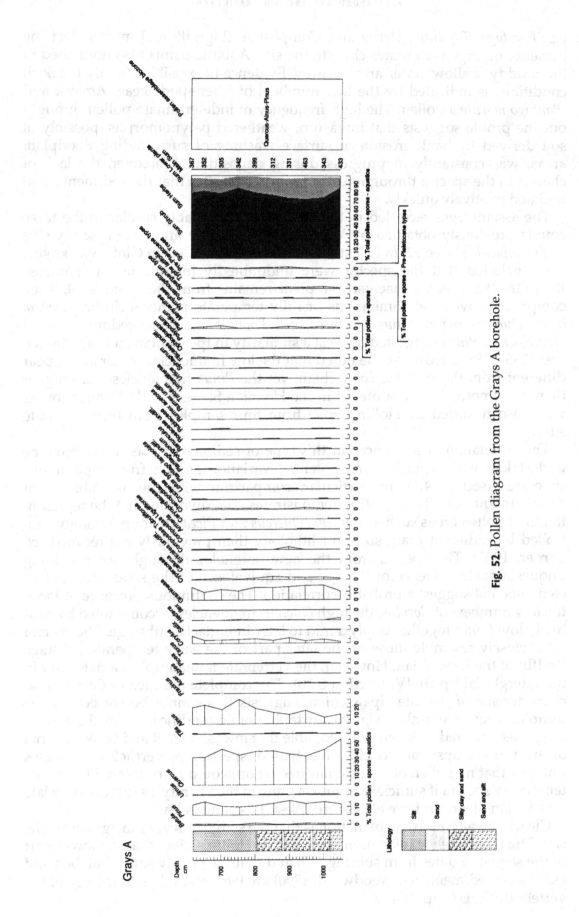

Fig. 52. Pollen diagram from the Grays A borehole.

e.g. *Plantago*, *Trifolium*, *Urtica* and Compositae (Liguliflorae), may reflect the presence of large vertebrates close to the site. Aquatic plants also flourished in the muddy shallow water areas nearby. Evidence of possible slightly brackish conditions is indicated by the low numbers of Chenopodiaceae, *Armeria* and *Plantago maritima* pollen. The high frequency of indeterminate pollen throughout the profile suggests that inwash of weathered palynomorphs, possibly in soil derived by bank erosion or surface washing of surrounding floodplain areas, was constantly in progress during deposition. Moreover, the lack of change in the spectra throughout the sequence implies that the sediment accumulated relatively quickly.

The assemblages recorded in this study are remarkably similar to the three counts previously obtained by West (1969) from grey and brown sandy silts with *Corbicula* above 9.5 m OD at the north end of the Celcon Globe Works site. He concluded that the spectra were undoubtedly interglacial in character. Reid (1897) recovered macroscopic plant remains from the same level. They comprised leaves and some seeds of fully temperate plants including *Hedera helix*, *Quercus robur*, *Alnus glutinosa* and *Populus* sp. The predominance of *Quercus* and *Pinus* suggested to West a similarity to Ipswichian substage IIb (cf. West 1957, 1980). However, he noted that the low frequencies of *Corylus* appear different from those of the Ipswichian, yet the *Pinus* frequencies were higher than those normally encountered in the Hoxnian Stage (Ho II). Similar counts were also recorded by Hollin (1977) from four samples from three separate sites.

The correlation of sequences in this type of sedimentary sequence must be undertaken with regard to the potential variation arising from taphonomic processes (section 4.1). In particular, comparison with the assemblages at Trafalgar Square (Gibbard, 1985) demonstrates unequivocally that the representation of pollen types such as *Corylus*, *Quercus* and *Picea* are more strongly controlled by sediment grain size and lithology than previously appreciated (cf. Turner, 1985). The spectra from the new borehole, although not providing unquestionable evidence of the age and equivalence of the sequence, contain elements that suggest a probable correlation. The continuous presence of *Picea*, the low numbers of *Corylus*, the high *Quercus* frequencies, accompanied by relatively low *Pinus*, together imply a mid to later interglacial substage. The spectra most closely resemble those of the later part of the early temperate substage (Ip IIb) of the Ipswichian. However, the vertebrate fauna implies a date later in the interglacial (Ip III–IV) (section 5.2b). The complete absence of *Carpinus*, so characteristic of the later Ipswichian, suggests this cannot be correct for the sampled sequence, unless significant taphonomic modification of the assemblage has occurred. It is certainly possible that inwash of soil and bank material derived from upstream could give biased spectra. Nevertheless it seems unlikely that no pollen of a particular tree taxon would be recovered from contemporary spectra if sufficient numbers grew in the vicinity, as indicated by late Ipswichian diagrams from elsewhere (West, 1980, and below).

Clearly, an alternative explanation is that the sediments vary in age across the site. The most likely explanation, however, is Hollin's view that the lower part of the sequence dates from substage IIb and that the pebbly sand shell bed and associated sediments (cf. Woodward, 1890) are later and relate to the age of the vertebrate finds (*c*. Ip III–IV).

6.2 West Thurrock

Two profiles have been analysed for their contained pollen and spores from West Thurrock clays in the road cutting (locality WTA) and the Lion Tramway Cutting (locality WTF) (section 2.7c). The pollen counts are shown in Fig. 53 and Table 2, respectively.

At WTA, although only four samples were analysed, two pollen assemblage biozones can be recognised:

0–75cm *Alnus–Carpinus* p.a.b. (Ip III)
75–90cm *Alnus–Pinus*–herb p.a.b. (Ip IV)

The spectra in the lower p.a.b. are dominated almost exclusively by the pollen of the temperate forest trees *Alnus* and *Carpinus*. Other tree taxa such as *Pinus* and *Corylus* are present only in insignificant numbers. Equally, the pollen of herbs, Gramineae and Cyperaceae represent less than 10% of the total assemblage throughout, implying that grassland did not occur near the site. This, together with the presence of Filicales and *Polypodium* spores, indicates that local alder fen carr was the dominant riverside vegetation. *Sparganium* also colonised the shallow muddy water nearby. The dry calcareous soils close by undoubtedly provided an ideal habitat for hornbeam forest.

The upper p.a.b. is represented by only one count. Nevertheless it shows a clear trend begun in the preceding biozone of decline of *Carpinus* and *Alnus* and their replacement by *Pinus* and *Betula*, accompanied by a low frequency of *Picea*, *Quercus* and *Corylus*. This marked change is also reflected in the pollen of Gramineae, Cyperaceae and herbs. The latter include particularly those of open, disturbed ground such as Compositae (Liguliflorae), Chenopodiaceae and *Plantago lanceolata*. The spread of local grassland, possibly replacing the carr vegetation, seems to be implied.

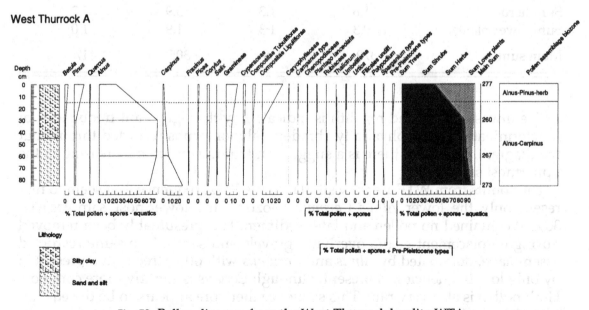

Fig. 53. Pollen diagram from the West Thurrock locality WTA.

Table 2. Pollen counts from the West Thurrock locality (WTF)

	F 30 cm	F 85 cm	F 110 cm	RGW
Betula	0.3		0.3	3.1
Pinus	5.2	0.4	2.6	6.7
Quercus				1.9
Alnus	88.8	64.9	67.2	52.7
Carpinus	1.9	24.9	18.5	25.4
Picea	0.8	0.2		0.2
Corylus	0.8	7.0	3.6	5.8
Salix	0.3			
Gramineae	1.1	0.9	4.6	1.7
Cyperaceae		0.4	0.7	
Compositae Tubuliflorae			0.3	0.2
Compositae Liguliflorae	0.5			0.2
Caryophyllaceae				0.2
Plantago lanceolata			0.3	
Ranunculus				0.2
Stellaria				0.2
Viscum				0.3
Sphagnum				0.2
Filicales undiff.	0.3	1.3	1.9	0.5
Polypodium				0.5
Nuphar				0.2
Potamogeton		0.4		
Sparganium type			0.6	0.2
Pre-Pleistocene types				0.5
Sum trees	97.0	90.4	88.6	90.0
Sum shrubs	1.1	7.0	3.6	5.8
Sum herbs	1.6	1.3	5.9	3.2
Sum lower plants	0.3	1.3	1.9	1.0
Main sum	379	459	305	417

The inwash of soil and regolith is indicated by the significant frequencies of indeterminate, mostly physically abraded palynomorphs recorded throughout the diagram. Moreover there is a suggestion that this input is increasing in the uppermost sample.

The counts from WTF are very similar to the WTA sequence, but seem to represent only the lower *Alnus–Carpinus* biozone; the lowermost samples (17, 30 cm) contained no pollen and later sediment had presumably been removed during emplacement of the overlying gravels and sands. The same restricted assemblage, dominated by *Alnus* and *Carpinus* with other tree taxa represented by only low frequencies, is present, although *Corylus* is slightly more common. Herb pollen is also very rare. This sequence therefore appears to be the equivalent of that at WTA.

The abundance of *Carpinus* in these sequences and its decline are characteristically found in the late Ipswichian in southern Britain. In this stage the change from substage Ip III to Ip IV is marked by a decrease from dominance of *Carpinus* and a rise of *Pinus* (Sparks & West, 1959; West, 1980). The *Alnus–Carpinus* p.a.b. and *Alnus–Pinus*–herb p.a.b. identified here probably correlate respectively with Ip III and Ip IV, as shown above. These ages therefore confirm Carreck's (1976) conclusion based on the vertebrate fauna, the single pollen spectrum obtained from the Lion Cutting by Hollin (1977) and an almost identical count from the same section by West (personal communication; Table 2).

6.3 Aveley, Sandy Lane Quarry

Pollen analyses from Aveley, Sandy Lane, were published by West (1969). The pollen diagram obtained is not illustrated here, but West's comments are summarised for comparative purposes. West obtained pollen and spores from two sampled profiles (his A and G localities), but countable samples were only found in the grey-brown silty clay mud (188–598 cm above the base) and the compressed mud with wood fragments (598–641 cm) in section A, as well as section G and the so-called '*Mammuthus* monolith'.

In this sequence West (1969) recognised two pollen assemblage biozones (p.a.b.):

Carpinus–Pinus–Corylus p.a.b. (Ip III)
Quercus–Pinus–Corylus p.a.b. (Ip IIb)

The spectra of the earlier biozone indicate temperate forest dominated by *Quercus*, *Pinus* and *Corylus* with other trees poorly represented. The lower frequencies of *Quercus* and *Corylus* in the basal part of the sequence indicate that the lowest sample was near the beginning of substage IIb. *Corylus* pollen declines towards the end of the biozone and this is accompanied by a rise in the pollen of *Carpinus*, *Tilia* and *Picea*. This rise is paralleled by an increase in the pollen of open-ground taxa indicating the development of herb-dominated weed communities locally. The colonisation of the locality by marginal reedswamp communities is shown by the upward increase in aquatic taxa and Cyperaceae in the detritus mud sediment.

In the *Carpinus–Pinus–Corylus* biozone, *Carpinus* had become an important forest component. *Quercus*, *Corylus* and *Tilia* decline, but *Pinus* remains significant. *Betula*, *Alnus* and *Picea* pollen frequencies all increase. Throughout the biozone frequencies of non-tree pollen (NAP) rise at the expense of tree pollen (AP), indicating the continued expansion of herb-dominated communities in the vicinity. This may reflect the growth of alluvial plains nearby. The considerable rise in *Salix*, Gramineae, Umbelliferae pollen and Filicales spores suggest the local spread of fen carr. Further expansion of open ground communities is indicated at the transition from detritus mud into the overlying clays. However, the silty clays lacked pollen and therefore the record could not be extended further. The overlying sediments can be presumed to represent later interglacial aggradation (cf. Hollin, 1977).

As shown above, West (1969) concluded that the two pollen assemblage biozones correlated respectively with substages Ip IIb and Ip III of the Ipswichian Stage, with which they closely correspond in terms of assemblage of taxa and relative arrival patterns. The zonal scheme at this site was attributed by West to successional rather than climatic change. Although there has been discussion of the relative age of this site (chapter 5), there can be virtually no doubt that it is correctly correlated in view of both the extraordinary similarity of the pollen sequence and the correspondence of individual pollen curves to the Trafalgar Square sequence (Gibbard, 1985). No doubt has been expressed about the correlation of the Trafalgar Square sediments with the Ipswichian stratotype at Bobbitshole, Ipswich (West, 1957); for example, see Franks (1960), Coope (1974) and Gibbard (1985). There is therefore no valid pollen biostratigraphical basis for accepting a different correlation for the Aveley Sandy Lane profile, bearing in mind that it represents a different sedimentary facies and a longer sequence. Any variations in fossil assemblages should therefore be primarily assessed for their environmental and taphonomic rather than their stratigraphical significance.

6.4 Purfleet

Four pollen counts were undertaken from the Greenlands Pit exposure by Hollin (1977) from his profile E. Well-preserved pollen was only found in the thick clay laminations. The counts all include high frequencies of *Quercus* and *Pinus* pollen, together with abundant *Alnus* pollen. *Corylus* is also common, but Gramineae and herbs are almost absent. Hollin states that 'this pollen belongs most probably to early Ipswichian Zone IIb, less probably to Hoxnian Zone II.' In the context of the other sequences described here the former indeed seems most likely.

D. T. Holyoak also collected samples from the Greenlands sections in 1981 (personal communication), and a single count from his site 1 (adjacent to the lane), at 37.5 cm above the base of the grey silt, is given in Table 3 . Unlike Hollin's counts, this shows a considerably lower frequency of *Alnus* and a higher frequency of Gramineae pollen. In other respects it closely resembles the earlier counts, except that it includes 1% of *Carpinus* pollen, a taxon lacking in all other counts from the site. These variations may suggest a slightly later date for Holyoak's count.

Two further counts were made by R. G. West (personal communication) from samples collected in May 1978 from the neighbouring Bluelands quarry section adjacent to the narrow lane. These counts from 20 cm and 50 cm (RGW section A: Table 3) above the base of the grey laminated clay and silt are both very similar to Hollin's and reinforce his conclusions.

Taking into account the sediment type, pollen preservation and sequence recovered, a broadly Ipswichian substage Ip IIb age seems highly probable for the laminated bed.

6.5 Belhus Park, Aveley

Pollen analyses from this site were undertaken from the western face of the

Table 3. *Pollen counts from Purfleet*

	Greenlands 37.5 cm above base (DTH)	A20 Bluelands (RGW)	A50 Bluelands (RGW)
Betula	4.0		1.5
Pinus	26.6	19.7	40.3
Ulmus	2.0	0.6	0.6
Quercus	12.8	33.8	8.1
Alnus	5.9	34.7	39.6
Carpinus	1.0		
Tilia		0.3	0.2
Picea		0.3	0.8
Abies		0.3	0.4
Corylus	1.0	4.0	4.5
Salix	2.0	0.3	
Gramineae	14.9	3.9	3.6
Cyperaceae	2.0		
Compositae Liguliflorae	2.0	0.3	0.2
Caryophyllaceae	2.0		
Chenopodiaceae	5.9	0.6	
Geranium	0.2	0.2	
Plantago lanceolata	1.0	0.3	
Rosaceae	1.0		
Umbelliferae	3.0		
Urtica		0.6	
Filicales undiff.	12.9	0.3	
Nymphaea	1.0		
Pre-Pleistocene types	6.5	0.3	0.2
Sum trees	52.3	89.7	91.5
Sum shrubs	3.0	4.3	4.5
Sum herbs	31.8	5.7	4.0
Sum lower plants	12.9	0.3	0.0
Main sum	102	332	472

Counts by D. T. Holyoak (DTH) and R. G. West (RGW).

M25 cutting excavations (section 2.7g) and are shown in the pollen diagram in Fig. 54. No previous analyses had been undertaken here.

A single *Quercus–Alnus–Pinus* pollen assemblage biozone is represented here. The spectra are dominated throughout by temperate deciduous tree pollen, of which *Quercus* and *Alnus* are the most important elements. They are accompanied by lower frequencies of *Acer* and to a lesser extent *Fraxinus* and *Tilia*. Of the conifers, only *Pinus* is well represented, although rare grains of *Picea* are also present at some levels. *Corylus* is notably present in low frequencies, whilst the continuous curve for *Hedera* confirms the temperate forest evidence from the tree

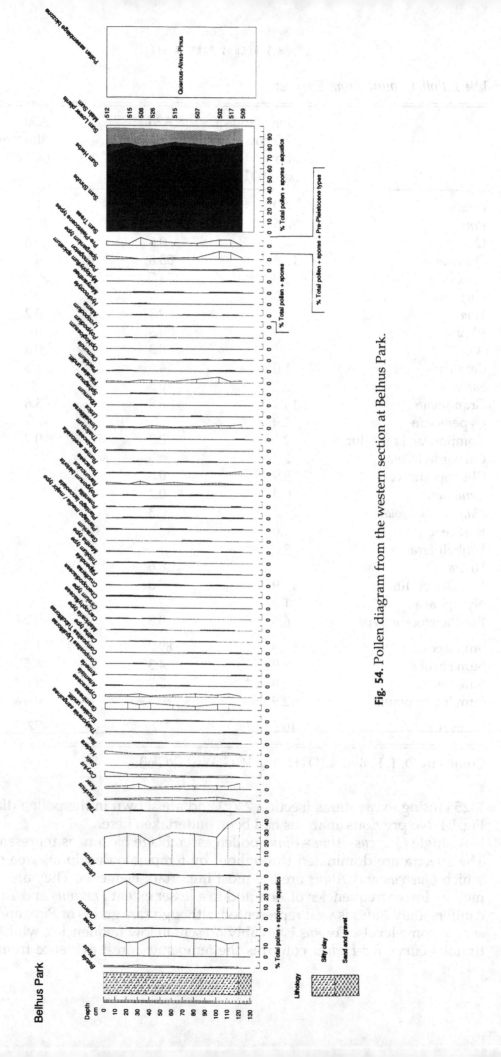

Fig. 54. Pollen diagram from the western section at Belhus Park.

pollen. A variety of herbs are represented, indicating the occurrence of local dry weed-dominated meadows, possibly on disturbed ground. These occur together with the pollen of plants of damp to muddy waterside areas. Of particular interest are the finds of low numbers of *Armeria maritima* and Chenopodiaceae pollen, possibly indicating the influence of saline water in the area. The local occurrence of damp woodland, possibly alder carr (cf. Phillips, 1974), is supported by the fern spores of Filicales and *Osmunda* recovered. The pollen of aquatic plants such as *Sparganium* and *Myriophyllum spicatum* is restricted to the lower part of the profile, possibly indicating that conditions developed from initial open water pool or channel to more closed marsh later. Indeed, the minor changes in the diagram appear only to reflect this development.

Overall, therefore, the vegetation indicated is of very local character, being that that would be expected to colonise an area adjacent to a river channel or floodplain pool, in which sediment rich in plant material accumulated. Dry ground beyond supported oak- and pine-dominated woodland. That inwash to the channel occurred constantly throughout the sequence is shown by the consistently low numbers of pre-Pleistocene types present.

Regarding the age of the sequence, the occurrence of high frequencies of *Quercus* and *Pinus* strongly suggests correlation with the first half of the Ipswichian Stage. The low frequency of *Corylus* and possibly *Acer* indicate that the profile should be equated with substage Ip IIa–b (cf. West, 1957, 1980).

6.6 North Ockendon

Samples from the light grey clay in the sections at Baldwin's Farm, North Ockendon, were pollen analysed and the results are shown in Table 4. The pollen and spores are well preserved in all but sample 260 cm, where they were too sparse to count. In addition, an unstratified sample of detritus mud from the excavations on Mr Mee's farm was also counted. All the assemblages recovered show very similar pollen spectra and will therefore be described together.

All the counts can be included in a *Pinus–Alnus–Quercus* p.a.b. They contain high frequencies of *Pinus* and *Alnus*, with subordinate *Quercus* pollen, other trees being insignificant. Apart from Gramineae and Cyperaceae, the pollen of herbaceous plants is rare, suggesting local weed-covered patches in an otherwise forested area. Of the forest, local stands of *Alder* carr would seem to have been most important, with pine growing on the better-drained gravels of neighbouring terraces.

These spectra closely resemble those obtained from Belhus Park (section 6.5) and therefore the comments regarding the age of the sequence are the same as those for the latter site. On this basis the assemblage should probably be equated with Ipswichian substages Ip IIa–b.

6.7 Ilford

The pollen sequence in Seven Kings borehole 30 in Ilford (section 2.7i) was published in detail by West *et al.* (1964). As at Aveley, the diagram is not reproduced here, but the main features of the sequence are presented for completeness.

Table 4. Pollen counts from Baldwin's Farm Quarry, North Ockendon

	2.30 m	2.90 m	3.20 m
Betula	0.9	0.7	0.4
Pinus	46.4	78.1	2.9
Ulmus	0.9		
Quercus	1.7	0.4	13.1
Alnus	36.1	2.8	62.0
Fraxinus			0.2
Acer			0.2
Picea			0.2
Corylus	0.9	0.4	1.0
Salix			0.2
Hedera			0.6
Gramineae	7.8	15.8	4.1
Cyperaceae	1.7		7.7
Ericales		1.4	
Compositae Tubuliflorae		0.4	
Caltha type			0.4
Caryophyllaceae			0.2
Filipendula			0.2
Lythrum			0.2
Plantago lanceolata	0.9		
Plantago media/major	0.9		
Ranunculus type			0.2
Rosaceae			0.6
Thalictrum			0.2
Umbelliferae			1.7
Flilcales undiff.	1.8		3.3
Pteridium			0.4
Potamogeton			0.4
Sparganium type			0.4
Typha latifolia			0.4
Sum trees	86.0	82.0	79.0
Sum shrubs	0.9	0.4	1.8
Sum herbs	11.3	17.6	15.5
Sum lower plants	1.8	0.0	3.7
Main sum	116	283	517

Five pollen assemblage biozones were identified:

563–653 cm *Quercus–Pinus–Corylus* p.a.b. (Zone *f*: Ip IIb)
653–660 cm *Pinus–Betula–Quercus* p.a.b. (Zone *e*: Ip IIa)
660–665 cm *Pinus–Betula* p.a.b. (Zone *d*: Ip Ib)
665–675 cm *Betula–Pinus* p.a.b. (Zone *c*: Ip Ia)
675–720 cm NAP–*Betula–Pinus–Juniperus* p.a.b. (Zone *b*: l Wo)

The basal Zone *b* comprises evidence for open, herb-dominated conditions with NAP frequencies of over 90% of total pollen. Very low numbers of *Betula*, *Juniperus* and *Pinus* pollen occur throughout, and suggest that scattered trees occurred in the neighbourhood. Gramineae and Cyperaceae are the most abundant herbaceous taxa, accompanied by a rich aquatic assemblage. The herb flora indicates a diverse aquatic flora that grew in and around the slow flowing stream or pond. Herb vegetation with few trees colonised the surrounding landscape, particularly the dry ground of the adjacent river terrace deposits.

The high NAP frequencies in the next Zone *c* indicate the continuation of herb domination. The increase in the *Betula* pollen is the main change. The Gramineae pollen remain common, but Cyperaceae and open-ground herbs decline, the latter possibly related to expansion of birch trees. A sharp increase in *Typha latifolia* pollen in the alternating organic and inorganic sediments suggests fluctuating water influx.

At the beginning of Zone *d*, the sediment becomes very organic. NAP levels decline and *Pinus* becomes more frequent than *Betula*, indicating the establishment of birch and pine forest at the expense of herbaceous vegetation. Gramineae numbers remain high in this zone.

Zone *e* sees the arrival of *Quercus* pollen, but this remains low throughout. *Pinus* is very common, but *Betula* numbers fall. The NAP is still abundant, with Gramineae and Cyperaceae the most frequent taxa. Aquatic types are limited to *Sparganium*-type and *Typha latifolia*.

At the start of Zone *f* numbers of NAP decline as forest became widespread. High frequencies of *Pinus*, *Quercus* and *Corylus* are recorded, with a range of other trees represented by low numbers. *Carpinus* and *Picea* frequencies increase at the end of the zone. Herbaceous pollen levels again rise in the middle of the zone, particularly of plants characteristic of waste ground. The sediments in which this zone occurs could be divided into a lower detritus mud and an upper clay mud. The latter is interpreted as reflecting flooding of the organic channel fill sediment by clay inwash. The detritus mud yielded macrofossils of several aquatic plants that indicate rising water level, such as *Potamogeton berchtoldii*, *Lemna* cf. *minor* and *Hydrocharis morsus-ranae*. Several herbs often associated with maritime conditions are also present (e.g. *Chenopodium* section *Pseudoblitum*, *Atriplex hastata/patula* and *Rumex maritimus*). The authors comment 'There is an indication of neighbouring maritime conditions in this assemblage and it is reasonable to explain the water level and subsequent flooding to the rise of sea level which is known to take place in zone *f* of this interglacial.'

Regarding the age of the sequence at Seven Kings, West *et al.* (1964) are in no doubt that it spans the latest Wolstonian to Ipswichian substage Ip IIb on the basis of the pollen assemblages recognised.

Exposures for the relief road south of Ilford High Road at Richmond Road were sampled by A. Currant (Section 2.7i). Here two samples from grey silt beneath the Ilford 'brickearth' yielded countable pollen and spores (Table 5). Sample A occurred 25 cm above sample C. Pollen preservation was poor in both.

The tree pollen predominantly comprises *Pinus*, present at a high frequency in sample C, but reduced in sample A, and low numbers of *Quercus*, *Alnus*, *Carpinus* and *Picea*. This appears to represent woodland, possibly some distance from the site on dry ground. Both samples contain very high frequencies of

Table 5. *Pollen counts from Richmond Road, Ilford*

	A	C
Betula	0.8	
Pinus	7.0	24.5
Ulmus	1.1	
Quercus	4.2	2.0
Alnus	0.5	0.4
Carpinus	1.1	1.2
Picea	0.5	1.2
Corylus	3.3	2.3
Salix		0.4
Solanum dolcamara		0.4
Gramineae	51.9	30.5
Cyperaceae	1.5	3.1
Calluna		0.4
Compositae Liguliflorae	3.9	14.3
Matricaria type	0.5	
Solidago type	0.8	2.3
Compositae Tubuliflorae	0.5	
Chenopodiaceae	0.3	0.4
Cruciferae	0.3	
Leguminosae undiff.		0.4
Lotus type	0.3	
Plantago lanceolata	16.9	10.7
Plantago media/major	1.6	1.9
Ranunculus	0.5	0.8
Rubiaceae	0.3	0.8
Rumex acetosa type		0.4
Thalictrum		0.4
Filicales undiff.	1.9	1.2
Polypodium	0.3	
Myriophyllum spicatum		0.4
Sparganium type		0.8
Indeterminable	12.6	10.1
Pediastrum	6.0	39.9
Pre-Pleistocene types	7.0	3.2
Sum trees	15.2	29.3
Sum shrubs	3.3	3.1
Sum herbs	79.3	66.4
Sum lower plants	2.2	1.2
Main sum	354	252

Gramineae and other herbaceous plant pollen, particularly of waste ground plants, and indeed they increase from sample C to A. A corresponding expansion of indeterminable pollen and pre-Quaternary microfossils suggests increased inwash of bank material that may be associated with the deforestation of the neighbouring ground. Aquatic plants are poorly represented.

The spectra obtained closely resemble those from Zone f in the Seven Kings borehole, in which *Carpinus* and *Picea* appear and *Corylus* and *Quercus* decline in the later part of the zone. On this basis the spectra are correlated with late substage Ip IIb of the Ipswichian Stage. As already mentioned, West *et al.* (1964) also note the accompanying expansion of herbaceous communities. This expansion of herb-dominated, often waste-ground assemblages on stream floodplain areas is a typical development of the second half of the Ipswichian Stage (e.g. Phillips, 1974; Gibbard & Stuart, 1975; Stuart, 1982). Here it may relate to floodplain aggradation resulting from sea-level rise, as well as possibly to the effects of vertebrates.

6.8 Nightingale Estate, Hackney

A series of samples was analysed from borehole NE2 at this site (Fig. 55). Pollen was well preserved throughout except in the uppermost sample where there were many corroded and degraded palynomorphs.

The sequence can be assigned to a *Quercus–Pinus–Alnus* p.a.b.

Tree pollen dominates the spectra at this site, the most important taxa being *Quercus*, *Pinus* and in the latter part *Alnus*. Other trees represented by low frequencies throughout include *Betula*, *Acer*, *Tilia* and *Ulmus*, whilst *Carpinus* and *Fraxinus* arrive in the upper part of the diagram. Together they indicate a diverse temperate woodland environment. *Corylus* and *Salix* pollen occur throughout. A diverse assemblage of herbaceous plants is represented, together with Gramineae and Cyperaceae. They indicate a range of habitats from damp marshy ground beside the channel to more distant dry grassland. Plants of dry waste places and trampled ground are well represented, particularly in the lower half of the sequence. The lower two-thirds of the sequence include pollen of a range of aquatic plants that grew in the muddy river channel.

The changes in the profile assemblages reflect local seral development rather than any climatic event. In the lowermost part of the profile considerable inwash of sediment is indicated by the very high frequencies of reworked pre-Pleistocene types. Also present in these spectra are significant quantities of pollen of herbaceous taxa, such as Compositae (Liguliflorae), possibly recruited by erosion of river bank material. The subsequent decline of the pre-Pleistocene types coincides with the change from light to dark grey silty clay and therefore presumably results from a marked decrease in inwash above. At this point the pollen of *Sparganium*-type rises, suggesting that still or slow flowing water was at least seasonally present.

The rise in *Alnus* pollen in the upper part, accompanied by an increase in Filicales spores once again, appears to indicate the colonisation of the immediate area by alder carr. This development is paralleled by a decline in aquatic taxa, tall herbs such as Umbelliferae, and *Pinus*, which together may reflect the spread of wetter conditions on the floodplain (i.e. local water-level rise).

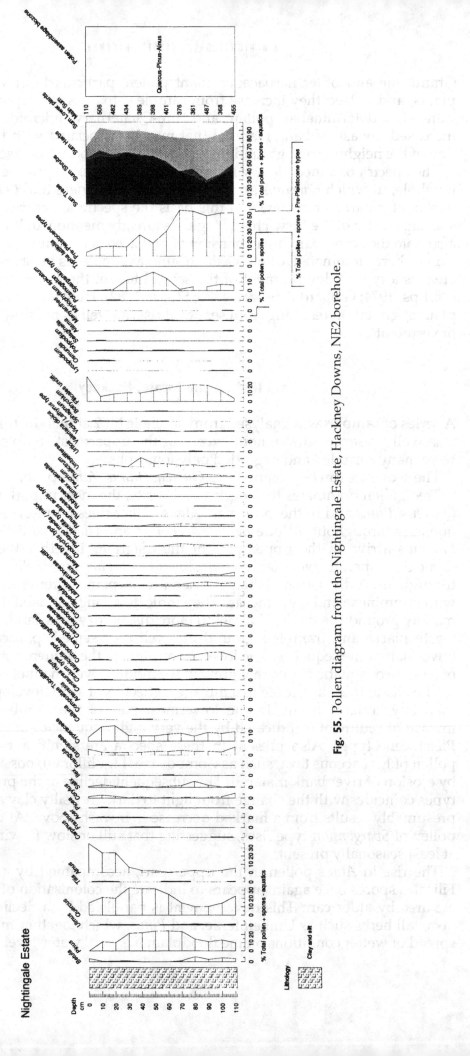

Fig. 55. Pollen diagram from the Nightingale Estate, Hackney Downs, NE2 borehole.

The age of the sequence based on the pollen spectra is similar to that of other sites discussed above. The predominance of *Quercus* and *Pinus* pollen suggests correlation with Ipswichian substage Ip II. The boundary between Ip IIa and IIb is conventionally placed at the expansion of *Corylus* (West, 1980). However, *Corylus* is present in low frequencies throughout this sequence therefore it is not possible to identify the boundary in the diagram. The marked increase in *Alnus* frequencies in the upper part of the profile may be masking a rise in *Corylus*, but the appearance of *Carpinus*, a taxon characteristic of the latter part of the temperate stage (Phillips, 1974; West, 1980; see also above), suggests that the entire sequence should be equated with substage Ip IIb.

Temperate macroscopic plant remains recovered from the nearby Shacklewell site of Prestwich (1855) by Reid (1897) show a strikingly similar assemblage including both *Quercus robur* and *Alnus glutinosa*. This strongly reinforces the lithostratigraphical correlation proposed here.

The Nightingale Estate sequence also shows remarkable similarity to those from the nearby Highbury locality (section 6.9). The sediments of these two sequences are lithostratigraphical equivalents (section 2.7j) and this correlation is confirmed by the biostratigraphy.

6.9 Highbury

As noted in section 2.7j, samples from two boreholes at St Augustine's Vicarage, Highbury, were analysed (Figs. 56, 57). Although pollen preservation was generally good, it was significantly poorer, being corroded and mechanically worn or broken, in the uppermost sample from both boreholes.

Both sequences comprise assemblages that can be assigned to a *Pinus–Quercus–Alnus* p.a.b.

The spectra here are again dominated by the pollen of trees, in particular by *Pinus*, *Quercus* and *Alnus* but with subordinate *Fraxinus*, *Tilia*, *Acer*, *Taxus* and *Picea*. Of the shrubs, *Corylus* is particularly important, but *Hedera* and *Salix* also commonly occur. The pollen of a range of herbs is present throughout, together with Gramineae and Cyperaceae. Dry calcareous grassland meadow plants are particularly well represented and suggest that some woodland clearings occurred in the vicinity. The pollen of marshy and damp ground plants is also present. By contrast, aquatic plant pollen is virtually absent from borehole 1, but present in the lower half of borehole 2, where the assemblage suggests muddy slow flowing or standing water. The occurrence of *Potamogeton* in the upper part of both sequences may hint that water level was rising at this time. This may have encouraged the expansion of wet woodland in low-lying adjacent areas, indicated by the considerable increase in *Alnus* and Filicales frequencies in the upper levels. The marked increase in *Alnus*, apparently at the expense of *Quercus*, suggests that this expansion was proceeding throughout the time represented. It presumably reflects a progressive rise in watertable, producing an increasingly wet, possibly more frequently inundated, floodplain.

At the same time a parallel decline in *Pinus* is seen. This tree probably inhabited the dry river terrace sediments adjacent to the channel. In the upper part of sequences, particularly borehole 2, it is partially replaced by *Taxus*. It is worth noting that the diagram from borehole 1 also includes 7% of *Carpinus* pollen in

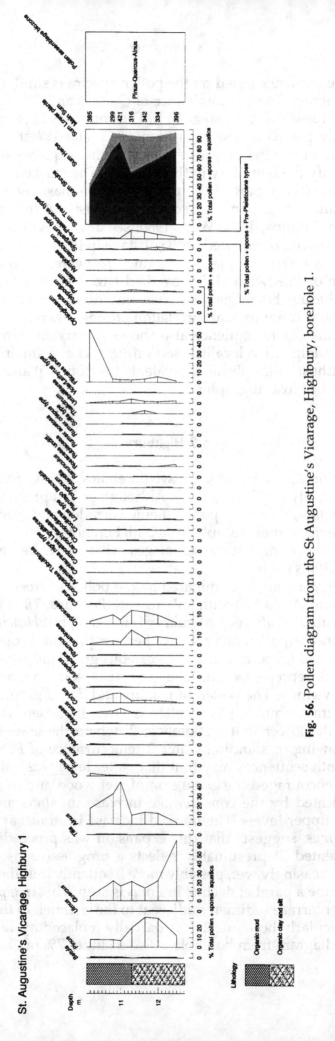

Fig. 56. Pollen diagram from the St Augustine's Vicarage, Highbury, borehole 1.

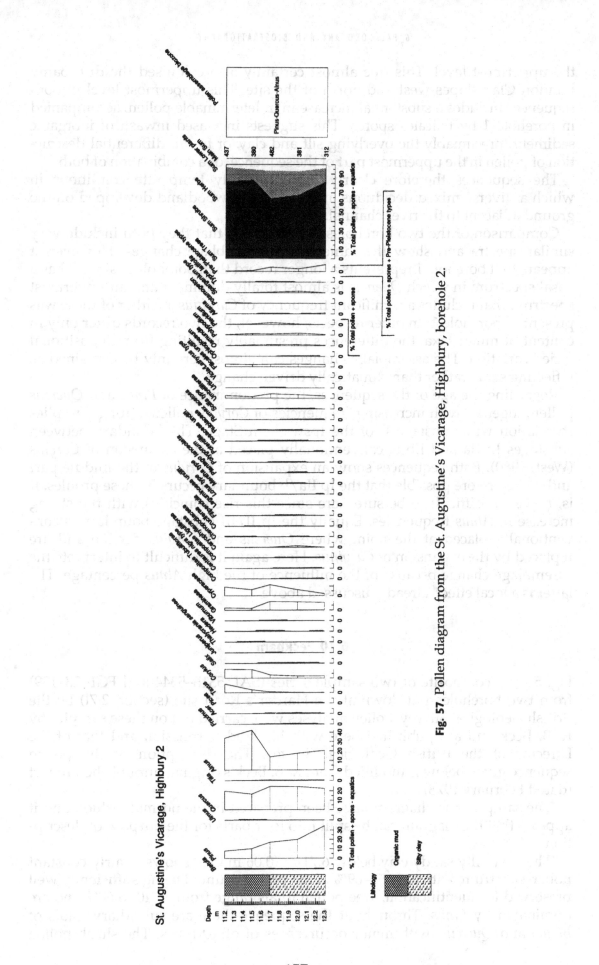

Fig. 57. Pollen diagram from the St. Augustine's Vicarage, Highbury, borehole 2.

the uppermost level. This tree almost certainly also colonised the dry loamy London Clay slopes west and north of the site. This uppermost level in both sequences includes a substantial increase in indeterminable pollen, accompanied in borehole 1 by Filicales spores. This suggests increased inwash of inorganic sediment, presumably the overlying silt and clay, or *in situ* differential destruction of pollen in the uppermost part of the sequence, or a combination of both.

The sequences therefore clearly represent fully temperate conditions in which a diverse mixed deciduous and coniferous woodland developed on the ground adjacent to the river channel.

Comparison of the two boreholes demonstrates that they both include very similar spectra and show the same broad assemblage changes. However, it appears that borehole 1 represents a longer record than borehole 2, since it has a basal spectrum in which *Quercus* is almost totally dominant and an uppermost spectrum that includes a significant frequency of *Carpinus*. Neither of these was present in borehole 1. In other respects, however, the two records differ only in content of minor taxa, the differences presumably resulting from depositional facies variation. The assemblage changes can almost certainly be explained as reflecting seral, rather than climatically driven change.

Regarding the age of the sequences, the predominance of *Pinus* and *Quercus* pollen, together with increasing frequencies of *Corylus* pollen, strongly implies correlation with substage II of the Ipswichian Stage. The boundary between substages Ip IIa and IIb is conventionally placed at the expansion of *Corylus* (West, 1980). Both sequences show an expansion of *Corylus* in the middle part and it is therefore possible that the Ip IIa/b boundary occurs in these profiles. It is, however, difficult to be sure here since this rise coincides with the strong increase in *Alnus* frequencies. Equally the Ip II/III substage boundary is conventionally placed at the point where *Quercus* and *Corylus* decline and are replaced by the expansion of *Carpinus*. Here again it is difficult to interpret the assemblage change because of the influence of the high *Alnus* percentage. The latter is a local effect, already discussed above.

6.10 Peckham

Fig. 58 is a composite of two sample series (SAL 5316–5344 and FGB 28–139) from two boreholes put down at the Harder's Road site (section 2.71) by the British Geological Survey. Pollen analyses were carried out on these samples by R. B. Beck and are published here with his kind permission and that of the Director of the British Geological Survey. The description of the pollen sequence given below is modified from R. B. Beck's original unpublished report (dated February 1978).

'The samples and diagram have been prepared in the normal fashion and it appears that the diagram can be split into four parts for the purpose of description.'

'The grey silty sandy clay below 6.71 m (0.06 m OD) shows a fairly constant pollen spectrum, with only 50–69% of the pollen counted being sufficiently well preserved for identification. Tree pollen totals range from … 40 to 60% and are dominated by *Pinus*. Throughout the sequence there are subsidiary totals of *Betula* and *Quercus*, with minor occurrences of other types. The shrub pollen

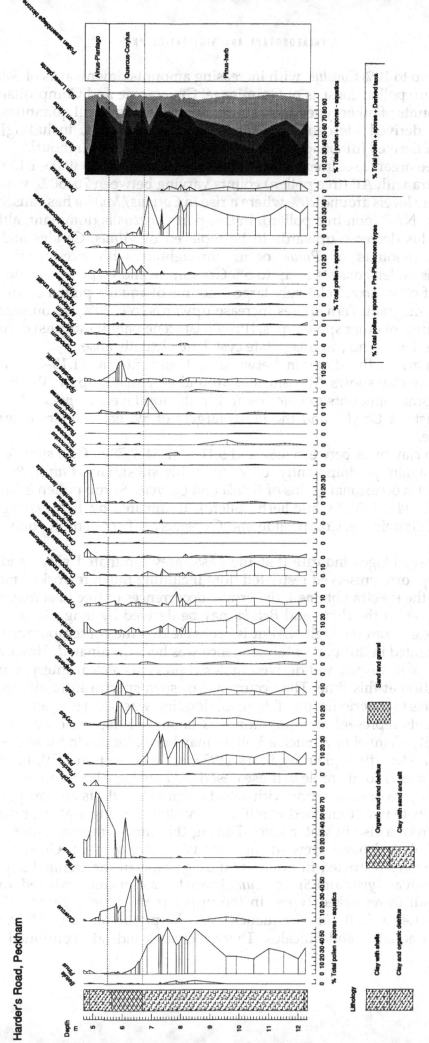

Fig. 58. Pollen diagram from the Harder's Road, Peckham borehole (counted by R. B. Beck).

comprises up to 10% *Corylus*, with increasing amounts downwards of *Salix* and *Ilex*. The herb pollen consists of Gramineae, Cyperaceae and Compositae, with minor amounts of other types. Fern spores are present in small quantities. High counts for derived Mesozoic and Tertiary [palynomorphs, including] large numbers of derived [dinoflagellate cysts are also present throughout].'

'The three organic sediment units between 5.64 and 6.71 m (0.06–1.13 m OD) show spectra with AP (tree pollen) counts varying between 70–50%, with three intermediate levels around 20% where a rise in *Corylus/Myrica* has caused a relative rise in NAP (non-tree pollen). In this part *Quercus* is dominant, although values of this decrease upwards to be replaced by *Alnus*, *Corylus* and *Salix*. Subsidiary amounts of *Pinus* occur throughout with occasional *Betula*. Herbaceous pollen consists of up to 5% Gramineae and Cyperaceae and minor numbers of other taxa. Relatively large amounts of aquatic pollen occur in this part of the diagram. Fern spores increase upwards to c. 20%, accompanied by small amounts of other spore types. [Pre-Pleistocene palynomorphs] occur only in the lower levels and [dinoflagellate cysts] are virtually absent.'

'In the part of the diagram between 5.11 and 5.64 m (1.13–2.34 m OD) ...the...peaty clay shows AP values of around 80% made up of 70–80% *Alnus* with very small amounts of *Quercus* and in the top level, *Carpinus*. The shrub NAP consists of *Corylus* and the herbs largely of 5% each of Gramineae and Cyperaceae.'

'The two clay units between 4.67 and 5.11 m (2.34–2.78 m OD) show 40% AP made up again predominantly of *Alnus* with subsidiary *Pinus*, *Picea* and *Carpinus*, with occasional grains of *Betula* and *Quercus*. Shrub pollen is low, consisting of *Corylus* (<5%) while herb pollen is dominated by 40% *Plantago* with subsidiary Gramineae, Compositae and Cyperaceae. Ferns are the only spores present.'

These assemblages indicate that the basal grey laminated silty sandy clay contains a poorly preserved restricted flora including much reworked material. In view of the spectra obtained, the erratic occurrences of tree taxa may also be reworked, whilst the *Pinus* and *Betula* may be derived by long distance transport. To judge from the high NAP it is probable that the contemporary vegetation represented in this part of the sequence was herb dominated. However, the high input of inwashed and derived material prevents a clear interpretation of the vegetation at this time. The occurrence of such spectra in laminated sediment implies that periodic, possibly nival flooding was occurring and therefore the sediments represent accumulation in a hollow or depression cut off from high velocity channel flow. Such a hollow may be analogous to those described from the modern floodplain by Berry (1979; see also section 2.71), in some of which sediments comparable to those described here have been found.

The change to organic muds with wood fragments reflects an abrupt change to sediments that accumulated in still or very slow flowing water, presumably resulting from a rise in water level. During this time temperate forest dominated by oak and hazel grew in the vicinity on dry ground. Grass meadow areas were very restricted in extent. Standing water in the channel supported aquatics such as *Typha* and *Sparganium*. Inwash was severely reduced, to judge from the fall in reworked types. In the upper part of the organic sediments above 6 m (1.45 m OD) *Alnus* frequencies rise sharply, accompanied by a similar increase in aquatics and Filicales. These increases indicate continued rise of water level, begun at the base of the unit, causing flooding of the local bank areas and concurrent retreat of dry land plants to higher ground. Sedimentary changes in this part of the profile reflect these water level changes.

The subsequent sedimentary change to clay is marked by a fall in the *Alnus* frequencies and a corresponding increase in herbaceous taxa such as Gramineae, Compositae and particularly *Plantago*. This influx of clay suggests renewed inwash of material derived by erosion of river banks and channels. The association of 'weed' type plants suggests that trees had retreated from the local river floodplain and were being replaced by meadow, possibly disturbed ground, vegetation. The development of similar grassland meadow floras is a particularly common phenomenon in the latter half of the Ipswichian Stage where it has been attributed to river floodplain aggradation and to a general opening of the forests (cf. Phillips, 1974; Gibbard & Stuart, 1975; Stuart, 1982).

Although the peat (2.01–2.34 m OD) was apparently radiocarbon-dated to >46 ka (catalogue number unknown), there can be little doubt that the upper part of this sequence represents a fully temperate environment since the spectra are dominated by deciduous forest trees. Beck states 'there is quite a close resemblance between levels 5.64–6.71 m and Zone *f* of the Ipswichian, provided that the high *Alnus* values are interpreted as a local vegetational phenomenon' (cf. also Berry, 1979). This accords well with other sites described above and with the Ipswichian stratotype sequence of West (1957, 1980). It is based on the predominance of *Quercus* and *Corylus* with subordinate *Pinus* in the assemblage, followed by the later arrival of *Carpinus* and *Picea*.

The change in composition of the assemblage in the overlying clay seems to indicate a sharp change of water level rather than a climatic event. This change is remarkably similar to that seen at Trafalgar Square (Tennessee Pancake House sequence) to the north (Gibbard, 1985), which is thought to be contemporary (cf. also Berry, 1979).

The age of the basal grey laminated clay is more problematic. Beck thought it 'possible that the correlation with the Ipswichian Zones d–e is most likely', although he found the high content of reworked pollen, particularly *Picea*, difficult to reconcile with this correlation. The author, however, considers that the combination of very high frequencies of reworked pollen and spores, possibly long-distance transported pollen and pre-Pleistocene palynomorphs, strongly suggest that the sequence accumulated under a cold climate during which there was substantial inwash of fines laden with reworked pollen and spores. Low pollen productivity and high sedimentation rates combine to cause local vegetation spectra to be swamped by the derived and/or long-distance transported component in such a situation (cf. Aario, 1947). In view of this, one is forced to conclude that the assemblage is of late-glacial type and predates the temperate sediments. By analogy with the sequence at Bobbitshole (West, 1957) and Seven Kings (West *et al.*, 1964, and above), this implies a late Wolstonian age (i.e. Zone a–b of Jessen & Milthers, 1928; l Wo of Turner & West, 1968; Phillips, 1974). An unconformity therefore occurs between these sediments and the overlying organic mud.

7
Palaeolithic artefact assemblages

Palaeolithic artefacts have been known from the gravels and sands of the Lower Thames area for 300 years. The first hand axe ever recorded was found by J. Conyers in 1690 in gravels beneath Granville Square, Grays Inn (Evans, 1860). The latter half of the nineteenth and first half of the twentieth centuries saw a great era of collection and recording of stone implements. This was the period of hand-working and sorting of gravel. Subsequently, the advent of machine operation has seen a decline in artefact discoveries, and indeed of vertebrate remains. It has also seen the exhaustion of artefact-bearing gravel resources.

The Lower Thames Valley is one of the most important areas for Palaeolithic artefacts in the country, a fact that may reflect its importance to Palaeolithic humans. This is because of the abundance of flint, the raw material for implement manufacture. In addition, the Thames has been a substantial river throughout its later history and would have offered an important routeway into the hinterland.

The artefact assemblages of the Thames Valley have been the subject of innumerable studies and important publications. The most famous local collectors included in the nineteenth century Evans, Spurrell and Smith, and in this century Smith, Chandler and Burchell and more recently Wymer and Sieveking. There are three recent synthetic works on the Thames assemblages: the *Gazeteer of Lower Palaeolithic sites* (Roe, 1968a) and the two important, detailed syntheses by Wymer (1968, 1985), already extensively referred to above. Wymer's work has been used as the basic reference and is the source of the data assembled in Table 6. It has been supplemented by reference to Callow (1976) and Roe (1981). The artefacts themselves have not been studied by the author. Detailed technological and typological analyses of artefact assemblages from a range of important Thames Valley localities and their relation to sites elsewhere are discussed by Roe (1981).

Attempts to reconstruct the stratigraphy and history of the terrace deposits using artefact assemblages and hand-axe typology have met with failure. The purpose of the work reported here is not to provide a detailed typological or cultural study, but to relate the assemblages that have been collected to the stratigraphy established above and to discuss some of the implications of the distributions.

The classification of pieces is based solely on the typological scheme of Wymer (1968, 1985) in which the following types have been recognised:

A Pebble chopper core (Clactonian)
B Biconical chopper core (Clactonian)
C Proto hand axe (Clactonian)
D Pointed hand axe (Acheulian)
E Pointed hand axe (Acheulian)
F Pointed hand axe (Acheulian)
G Subcordate hand axe (Acheulian)
H Cleaver (Acheulian)
J Cordate hand axe (Acheulian)
K Ovate hand axe (Acheulian)
L Segmental chopping tool (Acheulian)
M Ficron hand axe (Acheulian)
N Flat butted cordate hand axe (Acheulian)
   Blade (Levalloisian)
   Flake (Levalloisian)
   Tortoise core (Levalloisian)

The use of alphabet letters for each of the hand axe and related types has been modified, purely for convenience, so that transitional types classified as JK, for example, have been reclassified as J, or GH as G, etc. in Table 6. This method does not allow for the freshness of the individual artefact, which could not be included. Almost all the artefacts are composed of flint, but a few are of quartzite or Greensand chert.

7.1 Incorporation of Palaeolithic artefacts in sediments

Most Palaeolithic artefacts are recovered from the gravel member sediments that accumulated in the braided river environment under cold climates, as discussed above. Implements are also found to a lesser extent in the Silt Complex sediments ('brickearth') and occasionally in sand and associated finer sediments of temperate interglacial character. Archaeologists have long acknowledged that these implements must generally be in a derived state since some are heavily rolled. A quantitative scheme for assessing the degree of abrasion was offered by Shackley (1974). The amount of transport required to produce a rolled appearance from a fresh piece is not known, but may be quite considerable in view of the hardness of flint (for experimental results see Harding et al., 1985). However, there can be little doubt that where flakes, cores or broken hand axes from sediments can be rejoined, or where fresh material of various sizes is concentrated and includes flakes of small size, transport must have either been minimal or of low energy, which prevented abrasion of individual pieces.

It cannot necessarily be assumed that artefacts were produced immediately prior to burial in the sediments in which they are found today. Artefacts left or dropped on the ground may become incorporated into sediment by channel migration, overbank deposition on a floodplain or burial in aeolian or slope

7 PALAEOLITHIC ARTEFACT ASSEMBLAGES

Table 6. *Palaeolithic artefact assemblages from the Lower Thames Valley aggradations*

	Types	Dartford Heath Gravel	Swanscombe Lr Gravel	Swanscombe Lr Loam	Swanscombe Lr M Gravel	Swanscombe U M Gravel	Ingress Vale	Orsett Heath Gravel	Corbets Tey Gravel	Mucking Gravel	Northfleet	West Thurrock	Grays Thurrock & Orsett Road
Clactonian	A		+++		+	+							
	B		+++	++		+			+				
	C		+										
Acheulian	D				+	+			+	+			
	E			+	+++	+++	+	+	++	++		+	
	F				++	++	+	+	+++	+++	++		
	G				+	+	+		+		+		
	H								+				
	J						+		++	++		+	
	K	?+							+	+	++		
	L										+		
	M												
	N										+		
Levalloisian	Blade										+++		
	Flake									+	+++	++	
	Tortoise										+++	++	
	Core												

+, present; ++, frequent; +++, very common.

7.1 INCORPORATION OF PALAEOLITHIC ARTEFACTS IN SEDIMENTS

	Crayford brickearth gravel	Ilford	Stoke Newington floor gravel	Stamford Hill	Lower Clapton (Lea Gravel) basal gravel	Purfleet middle gravel, beneath Botany Pit	Purfleet top gravel	Wansant Pit Dartford	Bowman's Lodge	Lea Valley Gravel (Hedge Lane)	Kempton Park/ETM Gravel
						+	++				
	+					+++		+++			
					+	++	++				
		+	+	+	+	+				+	
		+++	+++	+++	+++	+					
		++	++	++	++	++			+	+	+
	+	++		++	++	+	++				
		+		+	+						
	+				+	+	+	++	+++	+	+
		+		+	+	+	+			+	+
											++
											+++
	+++	+			++		++	+++	+		
	+						+++				

deposits, particularly under a cold climate. As with vertebrate remains, implements may be recycled during later phases and redeposited in younger accumulations. Especially common in this respect is the formation of an implement-bearing gravel 'lag' deposit that is formed by removal of fines during high-velocity flow. This may particularly arise during channel migration when basal erosion occurs and or where solifluced regolith enters the channel.

It is possible that erosion of pre-existing implement-rich sediment could result in an apparently inverted sequence by the incorporation of artefact assemblages into younger sediments. This could occur because the upper levels of a sedimentary succession may be first removed, exposing progressively lower and therefore earlier sediment as erosion proceeds. Incorporation of indestructable components from eroded sediments might therefore begin with the youngest and end with the oldest material.

The more restricted occurrence of implements in finer sediments, particularly the Langley Silt Complex ('brickearth'), requires a different explanation. It is possible that industries represent working floors, i.e. the implements are undisturbed from where they were originally dropped and were subsequently buried by later sediment. This was the thesis advocated by Smith (1879, 1884, 1887, 1894) for north-east London. This origin is equally possible in floodplain overbank, overlap depositional situations. However, concentrations of artefact material could equally arise from sedimentary processes. On slopes, even of very low angle, large clasts (including artefactual material) may be rolled or carried over smooth silt or clay surfaces by currents of low velocity. This rolling would not produce discernible wear or abrasion of pieces, comparable to that arising from collision of gravel-sized clasts in a river channel. Moreover, solifluction or remobilisation of the artefact-bearing sediments on slopes can lead to their redeposition at lower elevations. If this material is carried to the valley floor, the assemblage may then be sorted and transported by the river. The potential for the recycling of artefacts is therefore very high in river valley situations.

The frequency of artefacts in river valley sediment implies that human activity was considerable and it may be that humans preferred to live in the valley environment. There are several reasons for thinking that this is correct. River valleys would have provided an ideal source of drinking water, flat land with lush vegetation even during cold periods in summer, and an ideal source of material for stone tool manufacture. In addition, rivers were almost certainly gathering places for large mammals that could have been captured for food. During temperate periods humans may well have avoided closed forest, because of the dangers from predatory animals and the difficulty of movement, in favour of more open floodplains beside rivers. There are in fact numerous well-documented associations of Acheulian industries with stream channels (Roe, 1981). However, the effect of concentration of material by river processes cannot be determined.

7.2 Relationship of the artefacts to the stratigraphical units

The assemblages of the artefacts from all the sites known to the author that could be related to the stratigraphical units defined above have been tabulated

(Table 6). Artefacts known to come from within a unit were assigned to that unit. Those from depositional boundaries (for example the interface between a gravel and an overlying silt bed) may belong to one or other deposit or the intervening period during which the surface was exposed. The data obtained vary greatly, some sites having produced only one artefact, whereas others have yielded hundreds. In order to compare the assemblages only simple presence/absence data supplemented by a simple indication of abundance has been used. This was selected in preference to complex statistical testing, such as that used in the Middle Thames region (Gibbard, 1985), because equally meaningful results were obtained using the simpler technique.

(a) Dartford Heath Gravel

With the exception of the Bowman's Lodge assemblage (see below), there is little evidence of artefacts from the gravels of this member. Confusion over the precise stratigraphical relationships of the gravels and sands, the overlying clayey silt channel fill and a lower unit described by Smith & Dewey (1914) at the Wansant Pit and associated localities may explain the few records that exist. However, Chandler & Leach (1908) could definitely attribute four pieces to this unit, including a small ovate hand axe. Wymer (1968) found no evidence for any Early Acheulian material.

As stated above, the Dartford Heath Gravel is the downstream equivalent of the Black Park Gravel of the Middle Thames region. The latter contains local concentrations of artefacts including both Clactonian and Late Middle Acheulian material, particularly in the Caversham Channel, but also elsewhere such as Hillingdon (Wymer, 1968; Gibbard, 1985). However, large stretches are apparently devoid of palaeoliths, so the limited number of finds from the Dartford Heath Member is not surprising.

The assemblage of material from the base of the younger, Bowman's Lodge clayey silt channel fill contrasts strongly with the few pieces from the gravel. The industry is very fresh, having not been transported, and appears to represent a buried land surface. It includes a large number of chopper cores (types A and B) as well as over 30 Acheulian hand axes and a variety of specialist tools. Also present are occasional Levalloisian flakes and tortoise cores (Wymer, 1968).

(b) Swanscombe Member deposits

The Swanscombe sequence is perhaps the most important Palaeolithic archaeological site in the country because it has yielded large numbers of palaeoliths and some fragments of hominid skull. This has meant that the sequence has been fully investigated and the stratigraphical distribution of the industries is well known. The assemblages are summarised in stratigraphical order.

I Lower Gravel The Lower Gravel at Swanscombe has yielded an abundant flint industry in mainly sharp to slightly rolled condition. The industry consists of chopper cores, flakes and flake tools all of Clactonian type (Ovey, 1964), with the possible exception of a hand axe (type E) figured by Smith & Dewey (1913). The artefacts are generally confined to certain horizons, particularly near the top but have also been found throughout. A few heavily rolled pieces are also known (cf. Wymer, 1968; Roe, 1981) from the lower part of the deposit.

2 Lower Loam This unit has yielded very few palaeoliths, except in its uppermost weathered part. Marston (1942) recovered four Clactonian chopper cores (type B) and some flakes from this level, mostly in very sharp condition. In the 1960s, Waechter (1970) demonstrated that some horizons in the unit contain artefacts, including conjoinable flakes, in very fresh condition. These scatters were interpreted as knapping areas representing periodic temporary drying of surfaces during accumulation of the silty sediments (Roe, 1981). Slightly rolled, scattered Clactonian material is also found on the surface of this unit (Waechter & Conway, 1977).

3 Lower Middle Gravel The flint industry from the sands and gravel of this unit is very abundant. It comprises typical pointed hand axes and characteristic flaking debris of Middle Acheulian type, all in slightly rolled condition and from a variety of levels in the sediments. Clearly the assemblage is transported, although the wear suggests that most has not moved far. The most frequent hand-axe form is type E and there are several examples of ficron shape (Roe, 1968b, 1981). A few Clactonian type pieces are also known from this unit and may possibly be reworked from the deposits beneath.

4 Upper Middle Gravel The palaeoliths in the Upper Middle Gravel do not differ greatly from those of the preceding unit. Apart from the find of human skull fragments (Marston, 1937; Wymer, 1968), the artefacts are again all slightly rolled, but are most common in the basal 0.5 m (according to Waechter & Conway, 1977). They are dominated by Middle Acheulian pointed hand axes.

5 Upper Loam and Upper Gravel The Upper Loam was shown by Marston (1937, 1942) to contain 'working floors' (apparently undisturbed spreads) of palaeoliths both in and at the base of the unit. These spreads are probably restricted, since they were not confirmed in later excavations (Roe, 1981). On the basis of taphonomic considerations (see above), it seems likely that they could have been moved and marshalled into localised groups by low-velocity currents, possibly local flood events. The palaeoliths are white patinated, are in mint or slightly rolled condition but include a few rolled pieces. The most frequent type is the flat ovate hand axes (type E) showing advanced features such as the *tranchet finish* (Roe, 1981), although a range of other, more pointed, types are recorded, together with some Levalloisian flakes.

Material from the soliflucted Upper Gravel is abraded and damaged and is thought to have been reworked from the Upper Loam (Wymer, 1968).

(c) Ingress Vale

The so-called shell bed at this locality has yielded some finely made, sharp, patinated hand axes of Acheulian affinity. Since Smith & Dewey (1914) considered that these sediments were the lateral equivalent of the Barnfield Lower Gravel, 'the records of highly evolved hand-axes were puzzling' (Wymer, 1968, p. 333). However, controlled excavation indicated that the artefacts were not from the shelly sand bed but from gravel and 'loam' filling a channel cut into this unit (cf. section 2.3). They therefore probably correspond to the Acheulian industry in the Middle Gravels, as Wymer (op. cit.) thought.

(d) Comment

The Barnfield Lower Gravel and Lower Loam Beds have therefore yielded exclusively a Clactonian industry. The overlying sands, gravels and related sediments contain an Acheulian industry, the appearance of which post-dates the Lower Gravel and Lower Loam. The Lower Loam and the upper part of the Lower Gravel are correlated with the Hoxnian Stage (Kerney, 1971; Wymer, 1974). This correlation is reinforced by the late Anglian–early Hoxnian age of the Clactonian industry at Clacton itself (Singer et al., 1973; Wymer, 1977). At Hoxne the Acheulian industry first appeared in the early temperate substage (West & McBurney, 1954), i.e later than the industry at Clacton. On this basis, the Clactonian industries were therefore placed in the late Anglian to early temperate substage of the Hoxnian.

Until very recently the Clactonian were considered to be the earliest industries in Britain. However, evidence from the Black Park Gravel in the Middle Thames (Caversham Channel and Hillingdon) and more particularly from the excavations at Boxgrove, Sussex (Roberts, 1986), have suggested that Acheulian hand-axe industries were present before the latest Anglian Clactonian. Therefore the significance of the stratigraphical dating of the arrival of the Acheulian in the Hoxnian is now less certain.

(e) Orsett Heath Gravel

Artefacts from this member are very rare. Those that have been recovered are almost all rolled and in poor condition. The exception is those from Craylands Lane site at Swanscombe where this member abuts the Barnfield Middle Gravels. Here slightly rolled or sharp ovate or cordate hand axes were reported by Smith & Dewey (1914). It is possible that they were collected from the Middle Gravels, which may have been exposed here, but more probably they come from Orsett Heath Member sands, having been derived from the Middle Gravels during downcutting. The other finds likewise probably represent reworking from earlier sediments.

This sparse record of Acheulian material corresponds exactly with that from the Middle Thames' equivalent Boyn Hill Gravel (cf. Gibbard, 1985).

(f) Corbets Tey Gravel

In contrast to the previous unit, the assemblages from the Corbets Tey Gravel are rich and generally fresher, although most are slightly rolled. The Middle Acheulian industries include sites such as St. Swithin's, Barkingside, Wanstead, Chadwell St.Mary and central London localities such as Gray's Inn Road and Stamford Hill (Lea: Stamford Hill Gravel). The Clactonian industry from gravel at Grays (Little) Thurrock is also included in this unit (Wymer, 1957, 1968). Of the Acheulian material, the most common type is the large ficron hand axe (type F) and cleavers (type H), neither of which are found in the Swanscombe deposits. It seems most likely that most of this material is reworked from pre-existing deposits or spreads.

The large assemblage of Clactonian and Acheulian palaeoliths from the Botany Pit at Purfleet should almost certainly be included in this unit. However, the occurrence of proto-Levalloisian cores and flakes suggests that a second, possibly younger industry might be present (cf. Wymer, 1968, p. 313) higher in the sequence. No other site in this member has produced Levallois implements.

The similarity of the assemblages from the Corbets Tey Member again exactly parallels the situation in the Middle Thames (cf. Gibbard, 1985) and reinforces the lithological and altitudinal correlations discussed in section 2.5.

The sediments at sites such as Greenlands and Bluelands Pits, Purfleet and the gravels overlying the organic sediments at Belhus Park, Aveley, have been interpreted as forming part of the Corbets Tey Gravel by Bridgland (1988a) and mapped by the BGS. However, stratigraphical evidence clearly indicates that all or part of the sediments represent post-Corbets Tey accumulations (see below) and therefore finds from these deposits have been excluded from this discussion.

(g) Mucking Gravel

Finds from the Mucking Gravel, the downstream equivalent of the Middle Thames Taplow Gravel, are more common than upstream. The reason for this is unclear. However, Palaeolithic assemblages have been found from this unit throughout much of East London, as far as Rainham. In north-east London, the important finds from the Mucking Gravel equivalent, the Lea Leytonstone Gravel, in the area south of Stoke Newington Common to Hackney Downs and Lower Clapton, include considerable quantities of artefacts, mostly in slightly rolled, rolled or very rolled condition. According to Smith (1894) the artefacts were present at two levels, the base of the gravel and the top. The assemblage closely resembles that from the Stamford Hill Gravel, immediately to the north, from which they were almost certainly derived. Similarly, the Leytonstone Gravels on the eastern side of the Lea Valley (section 2.6) have also yielded comparable finds. Further downstream a worn hand axe was collected from the gravels beneath the 'Crayford Brickearth'. In contrast, Clactonian artefacts have been found both in and on the surface of the gravel beneath the 'Grays Brickearth'. This gravel may possibly be part of the Mucking Member and if so this is the only locality in the area where these gravels contain Clactonian material. The artefacts are, however, thought to be derived from the Corbets Tey Gravel immediately adjacent to the lower units (Wymer, 1985).

In contrast, Levalloisian flakes have occasionally been reported from this unit, e.g. Wanstead, Hackney. It is possible that these may be stray finds from near the surface, from overlying colluvial sediment (e.g. 'brickearth') or are incorrectly provenanced, but if they are contemporary, this is the oldest unit from which they have been recovered.

Apart from these flakes, almost all the palaeoliths known from the Mucking Gravel are of Middle Acheulian culture and their condition indicates that they were derived from older sediments or spreads.

(h) Aveley/West Thurrock/Crayford Sands and Silts

The artefact assemblages from the fine sediment complexes included in this unit are described on a site by site basis.

I Northfleet The Baker's Hole locality is 'the most famous Levalloisian site in the Britain', according to Wymer (1968, p. 354). Most of the thousands of palaeoliths from this site appear to come from 'coombe rock' (soliflucted chalk bedrock) and are in mint, sharp or slightly rolled condition. This highly evolved industry includes flakes, the characteristic tortoise cores and blades. Associated with the industry are a number of hand axes in similar condition, including one *bout coupé* type N hand axe (cf. Roe, 1981). It seems likely from their occurrence that this material was swept downhill by the flow of the chalk regolith.

At Burchell's site and the British Museum sites, artefacts were principally derived from the basal gravel and sand (section 2.7a) immediately above the 'coombe rock'. This assemblage comprised an abundant Levalloisian flake industry typologically indistinguishable from that at Baker's Hole (Kerney & Sieveking, 1977). The overlying loess has also yielded Levallois pieces, whilst the temperate silts included large quantities of humanly fractured and worked bone. A few hand axes were also recovered from the Upper Loess bed above the temperate sediments (Wymer, 1968).

The sequence at Northfleet is crucial to the dating of the Levalloisian technique in Britain and is interpreted as spanning the period from late Wolstonian to early Devensian (Wymer, 1977; Kerney & Sieveking, 1977).

2 Grays Thurrock The occurrence of Clactonian material from the gravels underlying the 'Grays Brickearth' has already been mentioned. However, Clactonian artefacts have also been found in the 'brickearth', according to Conway (in Wymer, 1985). Indeed the same geologist reported that some artefacts were also found in a gravel unit overlying the 'brickearth'. Kennard (1904) previously reported a side-scraper from this same stratigraphical position. This gravel is probably the same as that reported by Tylor (1869) and correlated above with the West Thurrock Gravel (sections 2.7b, 2.8). It seems likely, in view of the occurrence of the Clactonian industry in the Corbets Tey Gravel immediately north of the 'brickearth' spread, that the pieces in the younger sediments are all derived from the older gravels.

3 West Thurrock According to Warren (1923, p. 607) 'the basement-gravel ... yields an abundant [Palaeolithic] industry characterised by ... 'tortoise cores' and Levallois flakes, and is closely succeeded in the overlying beds by the *Elephas primigenius* fauna.' He confirmed this observation in 1942, noting that a Levalloisian working floor lay at the foot of a buried cliff and was overlain by sand and loam. Warren's sequence descriptions correspond closely to those both by earlier workers and recent observations by the author (section 2.7c) in the area, particularly from the Lion Tramway section. The occurrence of an undisturbed knapping floor here indicates that the occupation predates deposition of the overlying sediments.

4 Crayford–Erith The Crayford sequence is famous for the two 'working floor' spreads discovered at the base of the 'lower brickearth' by Spurrell (1880) in Stoneham's Pit and later by Chandler (1916) in Rutter's Pit. Isolated pieces have also been found in the overlying 'lower brickearth' (Wymer, 1968). The industry is Levalloisian and includes blade-like flake material that could be refitted, and even a hammerstone. Broken, possibly butchered bones were associated with the implements, including a lower jaw of *Coelodonta antiquitatis*. Hand axes are very rare. Almost all the artefacts are in mint condition, the only rolled material probably coming from the gravel beneath (see above). As at West Thurrock, these 'working floors' indicate occupation of the gravel surface before emplacement of the overlying sediment.

5 Purfleet, Bluelands and Greenlands Pits Wymer (1985) considers that there are three industries at this important locality. The basal shelly gravel has yielded at least one rolled irregular side-scraper of Clactonian aspect. Similar material has been obtained from the so-called middle gravel, immediately above the laminated silts. From the former, Palmer (1975) also obtained unabraded Acheulian material. This led her to conclude that the original knapping area was close by. A Levallois flake was also found in the uppermost gravel.

6 Aveley, Belhus Park A small cleaver (type H), a tip of a broken hand axe and two type F hand axes have been recovered from the clayey gravel overlying the interglacial organic sediments at Belhus Park in the M25 motorway excavations (section 2.7g). These pieces were in sharp or slightly rolled condition. The significance of these finds is unclear. Similar industries are known from the underlying Corbets Tey Gravel and the artefacts could have been derived from this source. Their discovery is therefore of little chronological significance.

7 Ilford Very few palaeoliths are known from the sediments at Ilford. The only finds that can be unequivocally associated with the fossiliferous deposits are three Levalloisian flakes found by Corner (1903) in the 'lowest shell bed' (Hinton, 1900) and two by Johnson (1902), all from the Uphall site. All but one are unrolled.

8 Stoke Newington As already discussed in section 2.7j, hundreds of artefacts were collected from the deposits of the Stoke Newington area in the nineteenth century. Two 'Palaeolithic floors' were identified particularly by Smith (1894 etc.), although rolled and deeply stained implements have been recovered from at or near the base of the underlying gravel (see Mucking Gravel, section 7.2g). Less rolled and stained material occurred on the gravel surface. Some abraded material, as well as many sharp or mint, unabraded artefacts, came from the 'floor' itself, in the 'brickearth'. A high proportion of small hand axes and side-scrapers were found from the latter, but although Smith recognised a threefold subdivision into those from the three levels (i.e. the base of the gravel, the top of the gravel and the 'floor'), no systematic difference was found by Wymer (1968) on typological grounds.

In view of the geological sequence in which these industries occur, it seems highly likely that much, if not all, of this material has been transported downslope from Stamford Hill by periglacial slope processes, as already noted. This would explain the distribution of the artefacts, their apparent uniformity and their apparent lack of abrasion, except where they have been incorporated into fluvial sediments (as at the base of the gravel). Clearly they need not have been transported very far from their original site(s).

If this explanation is correct, then all that can be said of their age is that they are equivalent to or younger than the Stamford Hill Gravel (section 2.5 and above).

(i) Miscellaneous

A number of isolated finds have been made from the lower level gravel units of both the main river and tributaries. These are generally rolled or very rolled and are thought to have been derived from older, higher units, since little emerges from a systematic evaluation of the assemblages (cf. Wymer, 1968, 1985). Some significant finds have been made from the Lea Valley, however.

(j) Lea Valley Gravel

According to Wymer (1985), 'the archaeological evidence from these Devensian deposits is sparse and uninformative.' Warren (1912) considered that the rare finds of rolled stone tools were derived from older deposits. However, the discovery of two fresh Levalloisian flakes, one from the base of the gravel at Enfield Lock and the second from Ponders End, were thought to be contemporary by Warren (1938). A third artefact is a broken leaf point from Broxbourne. A second leaf point was also reported from gravel at Stratford. These are of Upper Palaeolithic type (see Wymer, 1985, for further discussion).

8
Palaeogeographical evolution of the Lower Thames Valley

From the evidence presented in the previous chapters, the sequence of events in the Lower Thames Valley during the Pleistocene can be reconstructed. The stratigraphical scheme based on all the evidence available is summarised in Table 7 and will be discussed in chronological succession. Where possible, palaeogeographical reconstructions for each of the members are shown in Fig. 59.

Unless otherwise stated, each aggradational unit is separated by a phase of valley downcutting, from which, by definition, no deposits are preserved. As already mentioned, abandoned upper surfaces of aggradational units have, in almost all cases, been modified either by later sedimentation or degradation. The most common of these modifications resulted from periglacial processes and soil development.

In addition to the evidence presented above, fossil Mollusca from several sites in the study area have been investigated for the shell amino-acid content for chronological purposes. The aminostratigraphical technique is based on the principle that isoleucine initially present in shell protein in the L-configuration epimerises over geological time to its non-protein diastomer D-alloisoleucine. The ratios measured from the proportions of these two acids have been used directly for correlation and relative age 'dating'. However, this correlation must be restricted to limited geographical areas that have experienced the same thermal history, since the epimerisation reaction is temperature dependent (Miller, Hollin & Andrews, 1979; Miller & Hare, 1980). For use as a 'dating' technique, ratios must be calibrated from sites of known age, i.e. those dated by other methods.

The geochronological significance of the 'dates' obtained by this technique is much debated, the results having been used by some authors (e.g. Bowen & Sykes, 1988; Bowen et al., 1989) to reorder the sequence independent of other evidence. Here the ratios determined from sites in the area will be mentioned in the discussion and are shown in Table 8. The acceptability of any individual aminostratigraphical 'age' determination as correct or or not is based here

Table 7. *Sequence of events in the Lower Thames (*in Middle Thames) Valley during the Pleistocene*

Event	Dating evidence	Climate	Stage
Aggradation of Tilbury Deposits	C14	t	Flandrian
Deposition of Shepperton Gravel and tributary valley floodplain gravel (c. 15 000–10 000 BP)	C14*	c	Late
Downcutting (c. 30 000–15 000 BP)			
Deposition of main mass of Langley Silts			
Deposition of East Tilbury Marshes Gravel and tributary equivalents (?c. 45 000–30 000 BP)	C14*	c–t–c	Middle Devensian
Downcutting			
Deposition of West Thurrock Gravel		c	?early
Downcutting			
Deposition of Aveley Silts and Sands and equivalents	p, v	t	Ipswichian
Deposition of Spring Gardens Gravel	———— ?c in part ————		
Downcutting			
Deposition of Mucking Gravel		c	
Downcutting			
Deposition of Corbets Tey Gravel		c	Wolstonian
Downcutting			
Deposition of Orsett Heath Gravel		c	
Downcutting			
Deposition of Swanscombe Middle Gravel	— m, v —— t–c ——		
Deposition of Swanscombe Lower Loam	m, v	t	Hoxnian
Deposition of Swanscombe Lower Gravel	— m, v —— c–t ——		
Downcutting			
Deposition of Dartford Heath Gravel		gl, c	late
Initiation of Lower Thames Valley and Glaciation		gl, c	Anglian
Deposition of Woodford Green Gravel and equivalents		c	?early
Downcutting			
Deposition of Buckhurst Hill Gravel and equivalents		c	
Downcutting			
Deposition of Debden Green Gravel and equivalents			pre-Anglian
Downcutting			
Deposition of High Beach Gravel and equivalents			

C14, radiocarbon dating; v, vertebrate remains; p, pollen; m, Mollusca; t, temperate; c, cold; a, aeolian activity; gl, glacial. The stage subdivisions follow Mitchell *et al.* (1973) and Gibbard & Turner (1990).

8 PALAEOGEOGRAPHICAL EVOLUTION

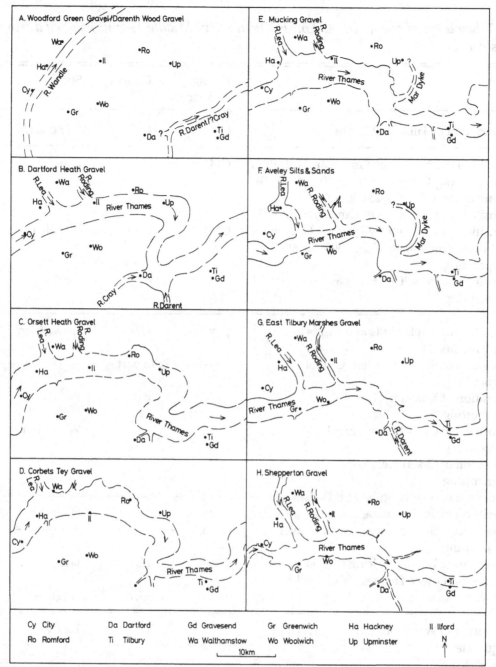

Fig. 59. Palaeogeographical reconstructions for the Lower Thames Valley area during the Pleistocene, based on the evidence presented here.

solely on the local and regional stratigraphy, using established litho- and biostratigraphical evidence. Few if any of the amino-acid determinations have been previously assessed in their full stratigraphical context. Seen in this light a proportion of the 'dates' appear to show overestimation of 'age' determined by conventional means. This suggests that variables other than those considered in the 'dating' procedures may be affecting the epimerisation reaction at some localities, such as groundwater geochemistry, temperature variation, burial, etc. On this basis it appears that the uncritical use of amino-acid epimerisation ratios for geochronological dating is premature. There is a need for detailed in-

Table 8. *D-Alloisoleucine/L-isoleucine ratios from non-marine Mollusca from Lower Thames sites (mean ratio and one standard deviation) from published sources*

Locality	Lab no.	Species	Mean, standard deviation	Reference source
Swanscombe: Ingress Vale	703	Cf	0.30 ± 0.017	1
	–	Cf	0.298	2
Barnfield Pit				
Lower Gravel	–	Bt	0.30 ± 0.015	2
Lower Loam	–	Cn	0.30 ± 0.01	2
L. Middle Gravel	–	Bt	0.296 ± 0.01	2
U. Middle Gravel	–	Bt	0.312 ± 0.017	2
Grays Thurrock ('brickearth')	702	Cf	0.31 ± 0.029	1
	–	Bt	0.29 ± 0.024	2
Crayford ('lower brickearth')	701	Cf	0.19 ± 0.024	2
	–	Cf	0.187 ± 0.007	1
	–	Bt	0.170	2
Aveley, Sandy Lane	811	Cf	0.19 ± 0.023	1
	–	B	0.148 ± 0.016	2
	–	Lp	0.172	2
Purfleet ('shelly gravel')	709	Cf	0.35 ± 0.032	1
	–	Cf	0.38 ± 0.07	2
	–	B	0.38 ± 0.24	2
('laminated beds')	710	Cf	0.36 ± 0.015	1
Ilford ('shelly bed')	700	Cf	0.23 ± 0.038	1
	–	Bt	0.23 ± 0.02	2
Clapton (unit unknown)	706	Cf	0.28 ± 0.036	1
Stoke Newington (unknown)	704	Cf	0.31 ± 0.026	1
Hackney (unknown)	705	Cf	0.27 ± 0.036	1
Trafalgar Square	–	Bt	0.11 ± 0.005	2
	–	Th	0.113 ± 0.005	2
	–	Cn	0.094 ± 0.004	2

Comparative ratios from British stratotype localities (source: 2)

		Aberystwyth lab	Boulder lab
Bobbitshole	(Ipswichian)	0.09 ± 0.01	0.1 ± 0.005
Hoxne	(Hoxnian)	0.261 ± 0.01	0.243 ± 0.023
West Runton	(Cromerian s.s.)	0.348 ± 0.01	0.346 ± 0.031

Molluscan species used: Cf = *Corbicula fluminalis*; Bt = *Bithynia tentaculata*; Cn = *Cepaea nemoralis*; Lp = *Lymnaea peregra*; Th = *Trichia hispida*.
References: 1: Miller, Hollin & Andrews (1979); 2: Bowen, Hughes, Sykes & Miller (1989).

and between-site investigations before the technique can be applied to 'dating' the type of complex sequences described here.

For this reason, the correlation of the stratigraphical sequence defined here with the global ocean record (e.g. Martinson *et al.*, 1987) using amino-acid determinations is not appropriate at the present state of knowledge.

8 PALAEOGEOGRAPHICAL EVOLUTION

8.1 Epping Forest Formation (pre-Anglian–?pre-glacial Anglian)

The highest and therefore the oldest fluvial deposits in the area are the highly dissected Epping Forest Formation gravel units. These gravels are of Weald origin and were deposited by the Wandle, aligned north-northeastwards across the present valley. This is based on the typical Weald-derived lithologies including Greensand chert, flint and subordinate quartz. The oldest unit, the High Beach Gravel, is represented both beneath Epping Forest and south of the river at Shooter's Hill. This is the earliest evidence of the Wandle. The next unit is the Debden Green Gravel. This can be equated with the Norwood Gravel by upstream extrapolation of the unit gradient. Similarly, the subsequent Woodford Green Gravel appears to be the downstream correlative of the Effra Gravel of the Crystal Palace area (Fig. 59). Reduction of the Weald pebble component in the Epping Forest area, in comparison with that in the area upstream, suggests incorporation of flint from local Tertiary rocks in the intervening area.

Evidence from the sedimentary facies preserved in the lower units suggests that all these gravels accumulated in a gravel bed river under cold climates.

8.2 Warley Gravel and Darenth Wood Gravel (pre-Anglian)

The Warley Gravel of the Brentwood area appears to represent a very early course of the Darent–Cray stream, aligned north or north-north-eastwards across the present Thames Valley to presumably join the Thames in Essex. The altitude of this unit suggests that it must equate with a member of the main river Pebble Gravel Formation or upper part of the Kesgrave Formation. As with the Epping Forest deposits, the pebble assemblage again typically comprises Kentish Weald materials, including the particularly characteristic Blackheath elongate flints.

By contrast, it has been shown that the lower level Darenth Wood Gravel would have been too low for the stream to have continued north of the present valley since the course was blocked by the Brentwood ridge. It is therefore thought that the contemporary Darent–Cray stream must have flowed eastwards, possibly along the approximate line of the present Thames course, to be confluent with the Medway (cf. Bridgland, 1988a; section 9.3) (Fig. 59).

This diversion of the lower course of the Darent–Cray stream must have taken place in the interval between the accumulation of the Warley and Darenth Wood gravels. The cause of this diversion is unknown, but the most likely explanation is capture of the Darent–Cray by a west-bank Medway tributary. The breach in the Chalk of the Purfleet anticline at Purfleet must presumably therefore be of considerable antiquity, since the Darent–Cray stream must have followed a course through this gap before its diversion. It is conceivable that once this northward-aligned lower course was abandoned, a west-bank tributary stream of the Darent–Cray could have occupied the gap and enlarged it at each successive downcutting phase.

8.3 Anglian Stage

Chalky till (Hornchurch Till) of typical Lowestoft Formation type is widespread on the northern side of the study area. The ice that deposited this till has been demonstrated to have moved into the area from the north-east and advanced over the pre-existing topography (cf. Allen, Cheshire & Whiteman, 1991). As it did so it advanced up river valleys, particularly that of the Wandle in the Loughton–Ongar area, where it overrode gravels and sands probably equivalent to the Woodford Green Gravel Member. Possible glaciolacustrine sediments intervene between the till and the fluviatile sediments at Chigwell and further north at Ongar (Baker & Jones, 1980; Baker, 1983). By analogy with the sequences in the Finchley and Watford areas (cf. Gibbard, 1977, 1979) they probably represent damming of the stream that culminated in drainage reversal, as previously proposed by Baker & Jones (1980) and Gibbard (1983). This also implies that the gravel and sand preserved beneath may represent the immediately pre-glacial accumulation of the pre-existing Wandle stream.

The Great Monk's Wood – Robin Hood's Pool Gravel possibly represents a glacially derived outwash, kame terrace-like accumulation deposited marginal to the ice tongue in the valley. If so, then the ice was at least 15–20 m thick. The gravels include an exotic assemblage derived from outside the London Basin. The water that deposited these sediments probably flowed southwards, since it is unlikely to have flowed towards the ice. This implies that the Wandle stream had been diverted and the Roding initiated by this time.

Elsewhere the ice seems to have been advancing close to its limit into the interfluve zone between the Wandle and the Darent–Cray catchment at Hornchurch. At the latter it apparently excavated a shallow valley-like depression in which till rich in London Clay was deposited. This depression cannot have been a normal pre-existing stream valley, since no fluvial deposits have ever been found beneath the till here.

The glaciation of the Vale of St Albans and the Finchley Valley, North London, caused diversion of the Thames and the Mole–Wey rivers by damming of their respective valleys and causing substantial ice-dammed lakes to form (Gibbard, 1974, 1977, 1979, 1983, 1985, 1989). Progressive overspill from the Vale of St Albans into the Mole–Wey valley and from there into the Wandle, across the intervening interfluves, caused a spillway system to develop. As already mentioned, an analogous ice-dammed lake may also have formed in the Wandle Valley. Clearly this development of spillways from valley to neighbouring valley will only have occurred until an unblocked valley was reached. By the Anglian, the Darent–Cray certainly was not aligned northwards parallel to the Wandle, north of the present Thames course, but eastwards to join the Medway. It therefore seems likely that water overflowing from the Wandle valley spilled into the Darent–Cray valley south of Crayford, possibly via a west-bank tributary. Incision of this spillway system caused adoption of the lower part of the Darent–Cray valley by the Thames. This would have carried the water eastwards into the undammed Medway near Southend-on-Sea, from where it flowed north to rejoin the original Thames' course near Clacton (e.g. Bridgland, 1980, 1988a).

The overspill of the water from the Wandle into the Darent–Cray valley would have required it to cut through or at least enlarge the pre-existing valley in the Chalk anticline at Purfleet. This ridge forms a substantial feature at present and might have been a barrier at this time. Younger deposits, the Orsett Heath and Corbets Tey members, adopt a very large meander immediately north of the ridge on the more easily eroded Tertiary rocks (sections 2.4, 2.5). There is no obvious explanation for this feature except that it possibly developed as the overflowing water attempted to break through the Chalk barrier or overtop the London Clay hills to the east. The course through the Chalk at Purfleet has remained a nodal point in the valley ever since (cf. Bridgland, 1988a).

On the basis of both the evidence assembled here and that previously obtained from the adjacent Middle Thames area, there is no doubt that the Dartford Heath Gravel, together with its upstream equivalent the Black Park Gravel, represents the first unit deposited by the Thames through the London area and the Lower Thames Valley. The marked change in lithology of the gravels, including exotic lithologies from the north (e.g. *Rhaxella* chert), the strikingly different trend of the river (almost perpendicular to the earlier alignment), and the occurrence of some Palaeolithic material all contrast substantially with the pre-existing situation. The Black Park Gravel has been shown to be of late Anglian age, on the basis of correlation with the Vale of St Albans' Smug Oak Gravel Member (Gibbard, 1979, 1985, 1989). By implication, the Dartford Heath Gravel is also of this age and therefore forms an extremely important marker unit throughout the Lower Thames Valley.

Detailed reconstruction of the course adopted by the Thames in the western part of the area is not possible because of the lack of deposits in East London. However, east of Aveley the course is well constrained. Fig. 59 illustrates the most probable course at this time. It is based in part on previous reconstructions by Gibbard (1979, 1985), as well as the new evidence from this investigation. The evidence available indicates that the Thames was confluent with the Darent and possibly the Cray at Dartford. North-bank tributaries seem to have developed as a response to drainage of the lakes in the respective valleys – the Roding occupying the lower Wandle (being analogous to the Brent occupying the lower Mole–Wey course) and the Colne occupying the Vale of St Albans Thames' course (cf. Gibbard, 1985).

An exception to this scheme may be the lower valley of the River Lea, which today extends from Ware due south to the Thames in East London. No unequivocal evidence of Anglian glacial deposits has been found in the Lea Valley, in spite of suggestions by some authors (e.g. Rose, Allen & Hey, 1976; Cheshire, 1981, 1983, 1986) that the valley existed. There is no doubt that glacial sediments occur in the Ware to Hoddesdon area, but these are certainly not associated with the Lea Valley as it exists today, but with the Mole–Wey valley that crosses the modern Lea here; i.e. the Lea Valley has been superimposed on this earlier valley fill. Near Hoddesdon gravels and sands overlie the Lowestoft Till (Cheshire, 1983) but this single outcrop is very difficult to relate to the Thames sequence. Nevertheless, it is possible that this unit represents outwash formed during the recession of the Lowestoft Till ice. If so, it could be a Lea Valley equivalent of the Black Park/Dartford Heath Gravel. The form and alignment of the Lea and Colne valleys are closely comparable and it is tempting therefore to conclude that they both originated at much the same time, i.e. the late Anglian. Both carry drainage from the abandoned Thames valley and neither can have existed before the glaciation.

Deposition of the Dartford Heath Member was followed by a substantial incision event. This incision, from the highest sediments of the Dartford Heath to the basal contact of the Swanscombe Lower Gravel Bed, represents a vertical change of level of a minimum of *c.* 14 m (allowing for the 8.25 km distance between the localities).

Accumulation of the lower part of the Swanscombe Lower Gravel has already been discussed by Bridgland *et al.* (1985) who concluded that the deposits represent a late cold climate, gravel-dominated river accumulation. The few fossils recovered include fallow deer, horse and aurochs. This fauna implies that the climate was mild, rather than severely cold. A few rolled Clactonian artefacts are also known. On the basis of their clast assemblage and stratigraphical position, there is no doubt that these deposits were laid down by the mainstream Thames in late Anglian time.

8.4 Hoxnian – early Wolstonian Stages

Deposition of the Swanscombe Lower Gravel Bed continued into the Hoxnian temperate Stage, as shown by its contained Mollusca and vertebrate fossils. Whether there was an hiatus between deposition of the lower and upper parts is unclear, but there is a marked decrease in the calibre of the material and an increase in the occurrence of fragile fossils. This implies a decrease in flow velocities and a colonisation of the channel, as well as the surrounding landscape, by plants and animals. According to Kerney (1971), both the Lower Gravel (upper part) and the succeeding Lower Loam contain an abundant freshwater mollusc fauna, 'of an association indicative of a large body of well-oxygenated, moving, calcareous water.' The occurrence of *Pisidium moitessierianum* is particularly important since it is characteristic of large rivers.

An overall trend occurs from the upper part of the Lower Gravel into and through the Lower Loam. This trend is seen as a change from the fine gravels, into a channel fill of silt with interbedded flood sand bands, the latter becoming rarer upwards, and finally to clay at the top. This gradual change from an open flowing channel to a dry floodplain surface is typically found in meandering streams in temperate regions that develop floodplains by vertical accretion (cf. Walker & Cant, 1984). This development of a floodplain was encouraged by a rise of the water level that caused flooding of the marginal ground recorded in the molluscan assemblages from the upper half of the Loam.

The terrestrial molluscan fauna from the Lower Gravel and Lower Loam represents dry woodland with dry grassland spreading in the uppermost part of the Loam. As mentioned above, a palaeosol is also developed in the upper part of the Loam. Formation of the soil, according to Kemp (1985), began on a dry surface, and leaching occurred during this period. It was later followed by development of a wet, gleyed soil that resulted from a second period of water level rise that may have been a precursor to the deposition of the overlying Middle Gravels. The period of palaeosol formation represents a considerable period during which a depositional hiatus occurred at Barnfield Pit, possibly equivalent to the second half of the temperate Stage. However, the deposits at Ingress Vale appear to represent the flowing channel facies deposited when the Lower Loam floodplain surface was exposed to subaerial weathering.

With the deposition of the widespread Lower Middle Gravel the river is rejuvenated and a return to more energetic, possibly more periodically 'peaked' flow is indicated. Local erosion of the Lower Loam surface was associated with the input of this material. However, erosion seems to have been very weak in places, since footprints are occasionally preserved on the surface. Nevertheless the occurrence of a basal 'lag' including vertebrate remains, Mollusca and Clactonian artefacts, derived from the underlying sediments, indicates that erosion of these sediments was taking place. Of the material thought to be contemporary with these and the overlying Upper Middle Gravels, the molluscan and small vertebrate evidence indicates that woodland was giving way to more open, herb-dominated grassland vegetation (?HoIV: Kerney, 1971). Of particular interest is the so-called Rhenish element of the molluscan fauna since it demonstrates (cf. Kennard, 1942a, b; Kerney, 1971) that the Thames and the Scheldt were confluent at this time Meijer (1988). This central European element is absent from the earlier units beneath. It therefore suggests that eustatic sea-level regression was in progress and was accompanied by the rivers extending their courses onto the newly emergent shelf areas.

Accumulation of the Lower Middle Gravel may have been followed locally by solutional bedrock collapse at Barnfield Pit.

Flowing channel-type deposits continued to accumulate at the base of the Upper Middle Gravel, the river flowing towards the ESE. The deposits comprise a gross fining-upward sequence, suggesting progressively reduced flow velocities as the area became infilled with sediment. The association of ground ice features within these sediments indicates periodic drying out of the area under a periglacial climate. The final deposition of the Upper Loam reflects the final abandonment of the local area by the river and its burial by colluvium from the adjacent slope.

Mollusca were rare from these beds, but the more common vertebrate fauna included grassland animals and fragments of a human skull, whilst the higher, 'Upper Gravel' yielded remains of the arctic musk ox. The human remains were recovered from about 60 cm above the base and were associated with numerous flakes and hand axes in sharp condition. Slightly rolled Acheulian hand axes were common in the pebbly units of the Lower and Upper Beds, whilst sharp hand axes and flakes were associated with the hominid skull, suggesting that Palaeolithic people were inhabiting the area immediately adjacent to the site. In addition, white patinated implements have been found immediately beneath the Upper Loam (Wymer, 1968).

The transition from late temperate into periglacial climatic conditions, indicated by the faunal assemblages, demonstrates that the Middle Gravel–Upper Loam sequence spans the latest Hoxnian/early Wolstonian Stage boundary (cf. Turner & West, 1968; West, 1980).

The geochronological evidence available for the Swanscombe sequence is conflicting. Two thermoluminescence (TL) dates obtained by Southgate (in Bridgland et al., 1985) from the Lower Loam (QTL 44A) and Upper Loam (QTL 44B) gave dates of 228.8 ± 23.3 ka BP and 202.0 ± 15.2 ka BP respectively. They place the deposits in Oxygen Isotope Stage 7 of Shackleton & Opdyke (1973). Similarly, provisional uranium series dating of bone from the Upper Middle Gravel by Szabo & Collins (1975) gave an age of >272 ka. By contrast, the amino-acid determinations on shells from the Lower Gravel, Lower Loam and the Middle Gravels give ratios of c. 0.3 that suggest an age equivalent to Oxygen Isotope Stage 11, i.e. c. 390 ka (Bowen et al.,1989, and Table 8). In comparison to the ratios of 0.24–0.26 determined for the equivalent stratotype Hoxnian sediments by both the Boulder and Aberystwyth laboratories, the ratios from the Swanscombe sediments would seem to be older. However, these determinations clearly conflict not only with the TL and U-series dates, but also with the litho- and biostratigraphy. It therefore seems that the amino-acid ratios are giving overestimates of the deposits' age at this site in comparison with those from Hoxne. The extensive decalcification of shells and calcite reprecipitation in the basal gravels observed here, or an unrelated geochemical condition, may be a factor in these determinations. Whether or not the TL and U-series dates are giving underestimates of age is difficult to decide. However, since the conventional stratigraphical position of the Swanscombe sequence is well constrained, the significance of all these geochronological estimations must await further evidence.

8.5 Wolstonian Stage*

Downcutting followed the infill of the valley at Swanscombe, and then aggradation of the Orsett Heath Gravel occurred throughout the Lower Thames Valley at virtually the same altitudinal range as the upper part of the Swanscombe sequence, as suggested by Oakley (in Ovey, 1964) and Gibbard (1985). The Orsett Heath Member also rests on the Anglian Hornchurch Till and, as demonstrated above, is the lateral equivalent of the Middle Thames' Boyn Hill Gravel. It is therefore of Wolstonian age. The gravels were deposited by the river in a braided form under a cold climate regime. The river's course during deposition of this unit is clearly aligned around the substantial Ockendon meander, north of Purfleet, as indicated by the outlier at Aveley (Fig. 59). No vertebrate fauna is known from this unit and finds of Palaeolithic artefact assemblages are very limited. The few that have been recovered are almost all rolled, are in poor condition and are thought to have been derived from the Swanscombe Middle Gravels during downcutting. The other finds likewise probably represent reworking from earlier sediments. This paucity of artefacts from the Orsett Heath Gravel corresponds exactly with that from the Middle Thames' equivalent Boyn Hill Member (cf. Gibbard, 1985).

The next youngest unit is the Corbets Tey Gravel. Once again this unit was deposited under a cold climate, with the river adopting a braided mode. The spreads of this unit are by far the most extensive in the area, including not only deposits of the main river, but also the wide confluence zone where the Lea and Roding joined the Thames in East London and neighbouring Essex (Fig. 59). The Thames continued to occupy the Ockendon meander at this time, and in the area adjacent to the Purfleet chalk ridge, solutional collapse of chalk bedrock occurred, leaving slope and pool sediments to interdigitate with the normal river sands and gravels. A Lea Valley equivalent of this member, the Stamford Hill Gravel, is also well represented.

Vertebrate faunal material is rare in the Corbets Tey Gravel, although mammoth, horse and bison, species typical of treeless environments, have been

---

* The term Wolstonian Stage is used here in the sense of Gibbard & Turner (1990) and represents all time from the end of the Hoxnian to the beginning of the Ipswichian stages.

recovered. The Palaeolithic artefact assemblages are rich and generally fresher, although most are slightly rolled. They include Middle Acheulian and Clactonian industries, the former dominated by the large ficron hand axes and cleavers. It seems most likely that most of this material is reworked from pre-existing deposits or spreads. In addition, the discovery of proto-Levalloisian cores and flakes suggests that a younger industry might be present higher in the sequence at Purfleet, since no other site in this member has produced Levallois implements.

The similarity of the Palaeolithic and vertebrate assemblages from the Corbets Tey Member again exactly parallels the situation in the Middle Thames (cf. Gibbard, 1985) and reinforces the lithological and altitudinal correlations discussed in section 2.5.

As with the previous gravel members, sedimentary evidence from the Mucking Gravel indicates deposition under cold climate, braided river conditions. This unit is well represented throughout the area and is the first to bypass the Ockendon meander. It again includes a substantial confluence area with the Lea and to a lesser extent with the Roding. The Lea Valley equivalent, here termed the Leytonstone Gravel, can be traced to the northern margin of the area studied.

Artefact finds from the Mucking Gravel, the downstream equivalent of the Middle Thames Taplow Gravel, are more common than upstream. The reason for this is unclear. Almost all the palaeoliths known from the Mucking Gravel are of Middle Acheulian culture and their condition indicates that they were probably derived from older sediments or spreads. For example, the important finds from the Lea Leytonstone Gravel in the Stoke Newington–Hackney area closely resemble those from the Stamford Hill Gravel, immediately to the north, from which they were almost certainly derived. Some Levalloisian flakes also occur in this unit and, if they are contemporary, this is the oldest unit in which they have been found.

The only vertebrate finds from the Mucking Member are elephant (?mammoth) and rhinoceros remains from Crayford. However, equivalent sediments in the Ebbsfleet valley include the treeless environment species mammoth *Mammuthus*, horse *Equus* and woolly rhinoceros *Coelodonta antiquitatis*.

In many places the Mucking Gravel and its equivalents are buried beneath younger accumulations, particularly the thick, fine estuarine, fluvial and colluvial sediment units of the Crayford, Ilford and Grays areas. It therefore clearly predates these fine sediments. It also predates the downcutting phase that immediately preceded deposition of the sediments at Aveley, Peckham and Trafalgar Square (cf. Gibbard, 1985). However, the Mucking Gravel post-dates the higher, Corbets Tey Member. It therefore appears to be of late Wolstonian age, in common with its upstream continuation, the Taplow Gravel.

At several localities such as Peckham, Ilford (Uphall Pit), North Ockendon, Aveley and Grays Thurrock, gravel and sand underlie the interglacial Aveley/West Thurrock/Crayford Silts and Sands complex. In contrast, at other sites, such as West Thurrock, Belhus Park, Purfleet and Ilford (High Street) and Nightingale Estate, a gravel 'lag' or older sediment unit appears to be present. The presence or absence of this underlying gravel and sand unit seems to reflect the position of the accumulation relative to the floor of the contemporary valley in which it occurs. This unit must have been deposited after a period of post-

Mucking Gravel downcutting in an interval before interglacial sediments began accumulating both in the main and tributary valleys including the Ockendon Channel. By analogy with other gravel units and because of its similarity in gradient, it may have been laid down under a cold climate. This unit is undoubtedly the equivalent of the Middle Thames' Spring Gardens Gravel, on the basis of its stratigraphical position (cf. Gibbard, 1985). Therefore it is proposed also to apply the name to this fragmentary spread in the Lower Thames.

Possible solutional collapse at Peckham produced an elongate depression that developed in the floodplain probably during deposition of the Spring Gardens Gravel. This apparently formed a lake-like hollow isolated from the main flow of the river, but into which flood water could penetrate, carrying with it fine detritus. The fine sediment infill thus produced records the contemporary, late-glacial type, herb-dominated vegetation.

On the basis of the sequence exposed at Northfleet in the Ebbsfleet valley, the late Wolstonian also included considerable colluvial activity, accompanied by contemporary aeolian sedimentation. Thermoluminescence dating indicates that the loess deposition occurred at $c.$ 150 ka (Wintle, 1982; Parks & Rendel, 1992).

8.6 Ipswichian Stage

Before any description of the events that occurred during the Ipswichian Stage in the Lower Thames Valley, it is important to stress that recently there has been considerable discussion over the age and significance of several of the interglacial temperate stage deposits in this area. As already stated (Chapter 5), this controversy has been based on the interpretation of individual sequences, particularly subtle variations in their contained vertebrate faunas, but also other fossil evidence, their altitude and geochronological determinations. Almost invariably these discussions have considered sequences in isolation, so that the lack of either a firm stratigraphical framework or an understanding of the sedimentary setting has led to considerable misunderstanding and confusion of the significance and distribution of certain parameters controlling the fossil assemblages etc. As shown in Chapters 5 and 7, there is strong biostratigraphical evidence indicating that the interglacial sediments at Northfleet, Grays Thurrock, West Thurrock, Crayford, Aveley, Purfleet, Belhus Park, Ockendon, Upminster, Ilford, Hackney, Highbury and Peckham represent parts of the Ipswichian Stage (*sensu* West, 1957, 1980). Equally, the lithostratigraphy (Chapter 2) and sedimentology (Chapter 4) reinforce the correlation. Therefore, unless it can be shown that the data conflict with the interpretation presented, there appear to be no strong grounds for considering a more complicated sequence. The sequence presented below represents that which emerges from a synthesis of the conclusions reached in this study.

The amino-acid racemisation data from Mollusca in the sediments correlated here with the Ipswichian Stage give D-alloisoleucine: L-isoleucine ratios that range from 0.38 (Purfleet) to 0.10 (Trafalgar Square) (Table 8). Comparison with the ratio of 0.10 from the stratotype Ipswichian at Bobbitshole suggests that either there is a range of ages represented, as supported by Bowen *et al.* (1989) or that there is a systematic error arising from local effects, already mentioned above. Assuming that the deposits are correctly correlated with the Ipswichian Stage, examination of the ratios obtained suggests broadly that higher altitude, often well-drained sites give higher ratios than those at low levels or in valley bottoms. Also, sites resting on Chalk bedrock seem to give higher ratios than those on Tertiary rocks. Nevertheless, there are large discrepancies that require explanation beyond the scope of this study. As these ratios suggest ages that conflict with established litho- and biostratigraphical evidence, they must remain suspect and are therefore not discussed further here.

No downcutting occurred at the end of the Wolstonian Stage, the transition into the Ipswichian being recorded by sediments at Ilford, for example. Here the fossil assemblages indicate an inward migration of boreal, then deciduous forest flora and fauna. By the early temperate substage, broadleaf, oak-dominated forest was established. The beginning of the late temperate substage is indicated by an increase in late temperate forest elements, particularly hornbeam *Carpinus* and to some extent spruce *Picea*. This change is accompanied by opening of the vegetation, with extensive grassland. In the post-temperate substage open vegetation continues to expand and finally boreal forests become widespread, such as seen at West Thurrock.

Following the late-glacial infill of the hollow at Peckham, there was a break in sedimentation. Subsequently, temperate sediments accumulated under still or very slow flowing freshwater in the floodplain depression or river channel. These sediments are of Ipswichian Ip IIb age. The sharp change to the overlying clays marks an abrupt inundation of the site, inwash of inorganic debris and an expansion of local meadow on the nearby floodplains. This sequence of events precisely parallels that recorded at Trafalgar Square 6 km to the north (see below).

Downstream in the main river the next site is Aveley, Sandy Lane. Here the thick valley-side sequence comprises stratified silts and clays, overlain by massive clays and capped by sand. The pollen stratigraphy clearly indicates that this sequence spans substages Ip II - III and this is supported by other fossil groups, particularly the vertebrates and Mollusca. Of particular interest here is the occurrence of the mollusc *Pseudamnicola confusa* (Holyoak, 1983). This indicates that saline conditions occurred during deposition of the silts at this site.

Nearby, at Purfleet, intertidal silts rest on a basal, shell-bearing gravel and reach a maximum height of 10.6 m OD. The gravel includes the freshwater molluscs *Corbicula fluminalis* and *Margaritifera auricularia* (Preece, 1988), indicating a fluvial origin for this unit. The marked sedimentary change to the overlying silts is also reflected in the fauna, which includes the brackish form of the marine ostracod *Cyprideis torosa*, together with some freshwater taxa (Robinson, in Hollin, 1977). This change appears to reflect marine transgression into this tributary valley. Since the pollen analyses from the silts indicate substage IIb of the Ipswichian Stage, the timing of the transgression would seem to correspond closely to that seen elsewhere.

The organic sediment at Belhus Park represents a channel that evolved from open water into a marsh. Initial water level rise was followed by the development of woodland in or beside the river channel. This sequence dates from Ipswichian substage IIa–b. Similarly the thick, channel-fill sequences at North Ockendon and Upminster appear to represent the same time and water level rise.

The sequence at Grays Thurrock apparently represents two parts of the

Ipswichian: the series of interbedded silts and sands encountered in the borehole date from a later part of Ip IIb, whilst the vertebrates indicate that later Ip III–IV sediments were also present. The interbedded silts and sands are almost certainly of tidal origin (cf. Hollin, 1977) and the presence of saline water is confirmed by the occurrence of the mollusc *Hydrobia ventrosa* (Woodward, 1890) and the ostracod *Cyprideis torosa* (Hinton & Kennard, 1901). The molluscan fauna also suggests that the basal gravel (beneath the silts) was of freshwater origin.

The complex of localities at Ilford represents the filling of a shallow tributary stream valley, almost certainly the Seven Kings Water and its extension to a confluence with the contemporary Roding, cut into the Mucking Gravel and local bedrock. At the Seven Kings locality, the sequence spans the period late Wolstonian to the first half of the Ipswichian. In contrast to the earlier freshwater molluscan assemblages, those from Zone *f* (Ip IIb) indicate a regional rise of water level and associated maritime conditions in the vicinity, according to West *et al.* (1964). Ip IIb deposits are also found beneath Ilford High Road. The Mollusca from Seven Kings suggest a small, sluggish, meandering stream with floodplains on the local gravels. They contrast with those from the High Road and Uphall sites which are characteristically river faunas. Of particular note is the occurrence of the brackish water species *Paladilhia radigueli* at Uphall at about 5 m OD (West *et al.*, 1964). Because the original sites in the Ilford area (High Road and Uphall quarry) are no longer available for investigation, it is not certain how much of the interglacial was represented by the sequences. Nevertheless Stuart (1978) concluded that the fully temperate faunal assemblage dated from the second half of the Ipswichian Stage (Ip III–IV).

On the opposite side of the valley from Aveley, Sandy Lane Quarry, lies the Crayford–Erith spread. Here the basal sediments are stratified silts of the fossiliferous 'lower brickearth'. The sediments and associated structures described from this unit strongly suggest that it is of tidal origin, as are those at Aveley, Grays and Purfleet. However, to the authors' knowledge no saline indicators have been found in the faunas. Indeed the occurrence of freshwater forms in the immediately overlying *Corbicula* bed indicate a recession of a marine influence by that time. Vertebrate remains from the 'lower brickearth' unit clearly indicate a late interglacial Ip IV to earliest Devensian age and that from the 'upper brickearth' again confirm sedimentary evidence for a return to true cold climates. The basal gravel at Crayford may have included a channel fill or 'lag' of temperate origin that predated the silts, since there are records of elephant and possibly hippopotamus from these gravels. These finds suggest an earlier Ipswichian (possibly Ip II) age for the 'lag'.

A remarkably similar series of apparently contemporary sediments is present at West Thurrock, banked against the valley side. Here a basal gravel 'lag' has yielded Middle Levalloisian artefacts. However, in contrast to Crayford, there is a thick sequence of current bedded sands with thin silts, overlain by a massive clay unit, the latter wedging into the sands from the south. The latter suggests a lateral facies change from rapid flowing water to still water conditions southwards from the valley side. The contemporary valley side here was a near vertical chalk cliff from which diamicton-like chalk clast lenses interfinger with the sands. These lenses suggest that the flowing water was eroding the cliff, causing material to collapse from time to time. The palaeontological assemblages from West Thurrock consistently indicate freshwater deposition. The vertebrate fauna from the sands is of later Ipswichian age, probably late Ip III–IV. Equally, the pollen from the possibly slightly younger clays indicates a transitional age from Ip III to IV.

The detailed investigations of the complex of sites at Northfleet in the Ebbsfleet Valley record the valley infilling. This infilling was initiated after deposition of the Wolstonian Orsett Heath Gravel, since the deposits rest in a valley excavated into this member. After downcutting by the Ebbsfleet, substantial quantities of colluvial material, predominantly slope wash and solifluction debris but with a loess component, accumulated at the base of the western valley side. The cessation in colluvial and loess deposition was followed by the establishment of a stable land surface under temperate (interglacial) conditions. A change to wetter conditions culminated in the reappearance of the stream at the site (Kemp, 1991), probably by inundation of its floodplain. This led to deposition of mollusc-bearing silts on the soil. This rise in water level is thought to reflect local watertable rise. Subsequent drying out is recorded by Kerney & Sieveking (1977), possibly reflecting a second period of stable land surface development and contemporary water level lowering. Ultimately, renewed colluvial and aeolian activity led to truncation of the underlying deposits and deposition of further solifluction and loess sediment under a cold climate.

On the basis of the temperate molluscan fauna and the thermoluminescence dates from the overlying and underlying 'brickearth' sediments at Northfleet, there is strong evidence that the temperate sediments and palaeosol at Burchell's site represent parts of the Ipswichian Stage. The underlying sediments are therefore Wolstonian, whilst those above are Devensian. In his description of the soil, Kemp (1991) remarks that the palaeosol he investigated was 'not particularly strongly developed'. This, coupled with the increasingly wetter conditions and followed by the silt deposition, suggests that the soil probably represents only a short period, possibly that before the substantial water level rise in the adjacent estuary caused water tables to rise locally (see below). If this is correct, then the soil may represent only substages Ip I–IIb, on the basis of the relative date of the water level rise from nearby localities. Subsequent water level fall seems to have initiated a second period of floodplain desiccation and soil development before deposition of the overlying colluvium.

The Mollusca from the temperate freshwater silts at Northfleet 'show no direct evidence of brackish water conditions, but their slack water facies arguably reflects estuarine sea-level of about 10 m OD' (Kerney & Sieveking 1977). Both the Mollusca and the vertebrate fauna suggest open, possibly treeless conditions near the end of the interglacial, possibly zone Ip IV. The characteristic Levalloisian flake industry from the lower 'brickearth', beneath the temperate sediments, is therefore of (late) Wolstonian age.

The Hackney Downs, Highbury, Shacklewell, Clapton and Stoke Newington area is underlain by a complex dissected spread of interglacial fluvial Hackney Downs Organic Deposits. The Nightingale Estate sequence records the fine sediment infilling of a channel-like complex under fully temperate deciduous forested conditions. Floodplains near the channel supported grassland that was trampled by vertebrates from time to time. Local water level rise, possibly indicated by colonisation of the floodplain by alder carr, is represented in the uppermost part of the profile. At neighbouring Shacklewell Lane, a virtually identical sequence was recorded but, in addition, yielded vertebrate material (?aurochs). The laterally equivalent sediments beneath Highbury represent a longer time period, but indicate a similar environment. Water level rise is also indicated by the increase in alder *Alnus* frequencies in the Highbury sequences.

The widespread sands that underlie the ground immediately north of Stoke Newington Common and continue east to Clapton Station, as well as south of Abney Park Cemetery and to Highbury, all appear to be related to a single fluvial aggradational event. The sands probably therefore reflect the deposition resulting from increased flow following the water level rise signalled in the pollen sequences. The deposits seem to represent an accumulation in a substantial meander bend by the River Lea, probably upstream of its confluence with the Thames. No evidence of a saline influence has been recorded from the sediments.

At both Nightingale Estate and Highbury the pollen assemblages obtained demonstrate that the sediments are again of Ipswichian age. The former correlates with late Ip IIb and the latter with Ip II to ?early Ip III. The inevitable conclusion from these results is that following initial deposition in shallow water, increased water level initiated aggradation across both the channel and floodplain areas during the latter part of the temperate Stage. This sediment was subsequently dissected, presumably early in the subsequent Devensian cold Stage, before deposition of the overlying Hackney Downs Gravel.

A comparison of the evidence from all the localities representing the Ipswichian Stage therefore indicates that almost all include substantial water level rise during Ip II and particularly in Ip IIb. That this rise reflects the eustatic sea-level rise known to have taken place at this time in the Ipswichian–Eemian Stage (West, 1972, 1980; Zagwijn, 1984) is repeatedly supported by the fossil assemblages and sedimentary facies.

The sedimentary facies association of the temperate deposits in the Lower Thames has been discussed and interpreted in section 4.3, where it was concluded that the lithofacies assemblage can be all accomodated within a river estuary. Indeed it is extremely clear that all the individual sediment sequences both in the main and tributary valleys can, without exception, all be explained by the model. Not only do the individual site sequences fit in terms of their lithofacies, but also in terms of their biostratigraphy and interpreted palaeoenvironments.

On the basis of the estuary model summarised in section 4.3b, individual site sequences would be expected to reflect the evolution of the river channel as the valley became progressively inundated, from a normal freshwater-type, fluvial sequence into the wide channel or flooded valley of the estuary itself. As the transgression proceeded, first the floodplains, then the bounding slopes and finally older terrace accumulations would have become submerged. Sedimentation on pre-existing deposits and surfaces would have led to their burial by the onlapping estuarine facies assemblage. Sedimentation would have been controlled by sea level, but the rate would be controlled by the volume of sediment supplied by the river. Therefore, once sea level had stabilised, both the main valley and tributary valleys would have progressively infilled up to at least mean sea level. Because of the complexities of water flow and water density contrasts in estuarine environments, it is normal for freshwater to dominate sedimentation in shallow areas and in zones of higher flow velocity.

Continued sedimentation would have caused the estuarine facies to migrate seawards; such a migration would happen under stable sea level and does not require marine regression. This would have led to progradation of entirely freshwater sediments both on and in channels cut into the immediately older estuarine deposits (section 4.3b). Ultimately, eustatically driven regression of the sea would be expected to have caused fluvial incision into the estuarine accumulations. This model is based closely upon the Flandrian evolution of the Gironde Estuary described by Jouanneau & Latouche (1981) and is indeed in many respects comparable to the Flandrian of the Lower Thames itself (cf. Devoy, 1979).

How well, therefore, do the individual Ipswichian site sequences outlined above relate to phases of development of the contemporary estuary? As Gibbard (1985) showed in the Middle Thames at Trafalgar Square, initial sedimentation took place in shallow floodplain channels or pools under freshwater conditions. Grass and herb-dominated floodplain vegetation was extensive adjacent to the channels and soils developed on the drier floodplain surfaces. This sedimentation pattern is precisely paralleled in tributary valleys, such as at Seven Kings, Northfleet and Highbury–Nightingale Estate.

However, as already repeatedly mentioned, evidence for subsequently increasing water depth or water table rise is seen in almost all the sequences examined from this time (substage Ip II). This rise initiated much more widespread sedimentation of silts and in some localities interbedded sand beds or larger sand units. For example, after deposition of freshwater fluvial sediments at Trafalgar Square, the increased water depth resulted in a clear separation of areas of high velocity flow and apparently relatively quiet water indicated by stratified sands and silts respectively. These sediments aggraded to almost 10 m OD. Nearby Peckham shows a remarkably parallel sequence. The sedimentation at sites such as Grays Thurrock and Aveley seems to have been initiated by this water level rise, because no pre-existing freshwater sediments appear to have been definitely present there. The sedimentary structures at these sites, particularly laminated inorganic sediments, suggest an intertidal origin, as suggested by Hollin (1977). Shallow water tidal sedimentation in the fluvial estuarine and estuarine zones seem to predominate in these sequences.

As Hollin (1977) also showed, the fossil assemblages, particularly the Mollusca but also Ostracoda, indicate that the rise in water level was related to eustatic sea-level rise during the interglacial. The invasion of the main and tributary valleys seems to have taken place during substage Ip IIb. The transgression apparently reached its maximum level of 10.6 m OD by the end of the substage, on the basis of the occurrence of intertidal silts at Purfleet. A remarkably similar figure is obtained from comparison of sequence thickness at other sites such as Aveley, Grays and Trafalgar Square (Fig. 29). Moreover, the maximum altitude of this transgression is constrained by those sites that occur above c. 11 m OD, such as Northfleet and Highbury–Nightingale Estate, which show no direct evidence of saline water incursion. Here sedimentation continued exclusively under freshwater conditions. Nevertheless, these sites do show substantial water table rise, such as might be expected as a response to transgression and accompanying sedimentation in the main valley. Saline wedges also penetrated up the substantial tributary valleys that were low enough to have been inundated, particularly the Mar Dyke valley (Ockendon Channel) in which Purfleet, Belhus Park, North Ockendon and Upminster experienced considerable aggradation, and the Roding, where brackish water reached Ilford. The water would have also overtopped the Seven Kings pollen sequence by 2.5 m at the maximum level. The increased alder pollen frequencies found at these sites closely parallel a similar development observed in the Flandrian sequences in the upstream part of the estuary by Devoy (1979).

It is interesting to note that the Ipswichian sequence preserved at Brentford (Gibbard, 1985) in West London also shows a transition from fossiliferous sands to brown silty clay 3 m thick between 7 and 11 m OD (Fig. 29). This clay could conceivably also reflect the water level rise in the estuary.

This high sea level obtained from the Lower Thames is somewhat higher than that from open coastal areas (West, 1972; Zagwijn, 1984). Setting aside any possible isostatic or tectonic considerations, it may be that the higher levels reflect the constraining and modifying effect of a funnel-shaped estuarine system on tidal activity. It is a commonly observed phenomenon that tidal water flowing into some estuaries is 'funnelled' and this causes a considerable rise of high water level or tidal amplitude depending on season, wind, etc. (Salomon & Allen, 1983). In the Thames system, tidal amplitude will almost certainly have been modified by this process, because the valley narrowed in a landward direction. According to LeFloch (1961), such a form is termed hypersynchronous and in such estuaries tidal currents attain maximum strength in middle or landward reaches.

As sedimentation filled the estuary, freshwater deposition progressively returned, beginning upstream and proceeding down both the main and tributary valleys. The later Ipswichian sequences such as those at Crayford and West Thurrock accumulated under these conditions. These sequences reflect the progradation of fluvial over estuarine sediments, mentioned in the model.

As has been shown above, all the Ipswichian sites in the Lower Thames Valley therefore indicate that a substantial estuary developed in the valley in response to rising eustatic sea level in the interglacial in substage Ip II (Fig. 59). The sea level stabilised at c. 10.6 m OD and the valleys became infilled. Subsequent progradation of fluvial sediments caused the sea to be progressively excluded from both the main and tributary valleys, so that by late Ip III freshwater deposits were once again being laid down in the Crayford–Grays area. On the basis of the West Thurrock sequence, it appears that sedimentation continued into substage Ip IV and possibly the earliest Devensian. No concrete eveidence for the timing of a marine regression can be determined from the sequences. However, incision of the sequences followed, presumably in the early Devensian (see below), to judge from the stratigraphical relationships in the Thurrock area.

8.7 Devensian Stage

Deposition of the West Thurrock Gravel followed the downcutting interval after final deposition and infill of the interglacial estuarine–fluvial complex deposits. The limited occurrence of this unit is difficult to explain, but probably reflects the intensity of subsequent fluvial incision that predates deposition of the lower, East Tilbury Marshes Gravel. The sedimentary facies are comparable to those of a braided river regime and therefore are probably of cold climate origin. It seems very likely that this unit should be assigned to a period post-Ipswichian and pre-Middle Devensian, i.e. early Devensian, on the basis of its statigraphical relationship to other units. If this is correctly dated, it invites comparison to the similarly aged, yet equally poorly represented Middle Thames' Reading Town Gravel Member (chapter 9).

The next youngest unit, East Tilbury Marshes Gravel, has already been shown to be the lateral equivalent of the Middle Thames' Kempton Park Member. Radiocarbon dates from the latter have indicated that it accumulated during the Middle Devensian substage, between 45 000 and 30 000 BP (Gibbard, 1985, 1989) (Fig.60). Unfortunately, no new dates were obtained during the present study. However, sedimentary structures are consistent from exposures in indicating braided river deposition. Unlike the Middle Thames area, few fossil finds have been made from these gravels. The only site of note is Greenwich, where a cold climate fauna has been recovered. Notably, this fauna included *Hippopotamus*, which was almost certainly reworked from local Ipswichian sediments and therefore emphasises the post-Ipswichian (i.e. Devensian) age of the unit.

Lea and Roding Valley equivalents have also been identified and termed respectively the Leyton and Redbridge gravels.

Chronologically, the East Tilbury Marshes Gravel and its equivalents are followed by deposition of part of the Langley Silt Complex. The latter is widespead in the London area but becomes relatively thin or absent to the east, such as in the area north of Tilbury. It is however, thicker and more common on the Kent side of the river valley, such as in the Swanscombe–Northfleet area. The Silt Complex sediments are predominantly found in positions adjacent to steep valley sides, often cloaking older accumulations. It is clear from the descriptions, mode of occurrence and analytical results that these sediments are polygenetic. Much of the sediment appears to comprise colluvial debris reworked from Tertiary clays and/or pre-existing Quaternary sediment. Some of the deposits include high frequencies of silt that might be of aeolian origin. However, other origins cannot be excluded. For example, it is possible that some material in appropriate positions on older units may represent weathered and partially reworked estuarine sediments that might have been expected to blanket older sediments in positions where they would have been drowned during high sea level stands. Overall, therefore, it is obvious that this sediment is of multiple origins, much of it representing local colluvial (solifluction and slope wash) deposition. Its conflicting stratigraphical occurrence also shows that it is certainly polygenetic and has originated at several different times.

Gravel and sand units beneath the floodplain of the River Lea have been shown to contain cold climate or 'full glacial' plant assemblages and vertebrate remains, referred to collectively as the Lea Valley Arctic Bed. Radiocarbon dates indicate that these sediments range in age from 28 000 to 21 000 BP.

The sediments containing the 'Arctic Bed' occur beneath a low terrace 1–2 m above the modern floodplain surface. The 'terrace' surface extends from Broxbourne to Tottenham, on the western side of the river. However, from the descriptions, it seems that the underlying deposits are continuous, at least in part, with the sediments beneath the alluvium. Nevertheless, the gravel and sand accumulation associated with the 'Ponder's End Stage' must predate the final 'buried channel' gravel unit (see below). However, neither the previous descriptions nor modern observations offer clear evidence for an erosional event between these two accumulations. Therefore, the term Lea Valley Gravel is used here for the gravel and sand unit underlying both the modern floodplain and Warren's 'Low Terrace'.

Rather than representing a single aggradational unit, it seems more probable that the 'Arctic Bed' was formed as a series of floodplain channel or depression fills. If this is so then the 'Low Terrace' may represent an accumulation in the wider part of the valley that escaped later modification.

Downcutting and incision to the lowermost level by the Thames (the 'buried channel') was followed by aggradation of the Shepperton Gravel. These events are paralleled by equivalent events in tributary valleys including the Fleet, Lea, Roding, Darent and Cray. The Shepperton Gravel almost totally comprises flint accompanied by only very minor quantities of other lithologies, the elimination of which progressed from the Dartford Heath Gravel to this member. Nowhere in the area is the Shepperton Gravel exposed, but tributary equivalents are disposed into sedimentary facies comparable to those of earlier units that indicate braided river deposition. Evidence from upstream (Gibbard, 1985) indicated that this deposition took place under a cold, probably periglacial climate between 15 000 and 10 000 BP. This was therefore the last major gravel and sand aggradation in the Thames system (Fig. 59). The dry valley systems present throughout the area appear to grade into this or equivalent aggradations, suggesting that they last functioned regularly during this interval.

The occurrence of collapse features beneath the gravels and sands resembles those found upstream (Gibbard, 1985) and in older units in the area. The situation of these structures suggests that syndepositional Chalk bedrock solution has been active under a periglacial climate during accumulation of the sediments. Larger scale collapse features beneath central London may have arisen from solution resulting from upwelling of artesian springs, enlarged by fluvial scour during gravel emplacement. Further discussion of these features is given by Hutchinson (1991). Contemporary collapse of gravel and sand into these hollows has in some places continued into the Flandrian and underlines that periglacial conditions need not be invoked to explain their operation.

8.8 Flandrian Stage

Climatic amelioration at the end of the Devensian was apparently followed by the river ceasing to aggrade sand and gravel and becoming stabilised within the flow channels it occupied at the time. This stabilisation was undoubtedly a consequence of the marked reduction in spring flooding, arising from mild winters, a drastically reduced sediment supply brought about by colonisation of the land by lush forest on valley sides as well as on the floodplain itself, and lastly by ground-water rather than surface-water drainage following the decay of permafrost. This stabilisation of the ground surface led to the river receiving very small amounts of sediment. With the development of woodland on the floodplain, peat began forming in pools and moist depressions. As this sediment thickened, it progressively buried the relict braided river surface developed on the Shepperton Member.

This surface is overlain by a complex of Flandrian sediments that comprises an interbedded sequence of alluvial sediments that dip and thicken markedly from west to east, collectively referred to as the Tilbury Deposits. According to Devoy (1979), the inorganic clay and silt units were deposited in a brackish water environment under tidal influence in the estuary. These units indicate the occurrence of four or possibly five marine transgressions, termed Thames I–V by Devoy (1979). Interbedded biogenic sediments mostly comprise monocot or wood peats and associated gyttja. At the eastern end, saltmarsh peats dominate but grade upstream into freshwater wood and fen peat. This facies is taken to indicate relative regression phases (Tilbury I–V).

The history of events in the Lower Thames caused by the Flandrian eustatic marine transgression has been investigated in detail by Devoy (1979). Only the main events will be summarised here.

The record begins at 8500–7000 BP, during which a rapid sea level rise from −25.5 to −8.9 m OD occurred (Devoy's Thames I phase). This was followed by a regression (Tilbury I) that lasted from 7000 to 6700 BP, indicated by the formation of alder wood peat. The Thames II transgression began at 6600 and continued to 5500 BP, during which sea level rose from −10.1 to −5.0 m OD. From 5500 to 4000 BP a substantial regression (Tilbury II) is recorded. Thick monocotyledonous peat accumulated downstream, passing into wood peat upstream at this time. A further rapid rise in sea level took place between 4000 and 2800 BP (Thames III) to −1.4 m OD. After a minor regression the mean sea level reached 0.4 m OD at Tilbury in *c.* 1750 BP. Recent continued rise (Thames V) may reflect anthropogenic modification of the estuary.

9
Correlation with neighbouring areas

The continuity of the fluvial aggradations recognised throughout the Lower Thames Valley and in some tributary valleys means that equivalent deposits must exist in neighbouring upstream and downstream areas. Upstream correlations have already been extensively discussed in earlier chapters, but here a summary of stratigraphical correlates in adjacent areas will be presented (Table 9). The correlations are made on the assumption that differential crustal warping has not been demonstrated between the areas.

9.1 Central Essex

The Kesgrave Formation ('Group') deposits are aligned approximately north-east to east across central Essex. They represent buried ancestral Thames sediments that were laid down before the rivers' diversion (Rose & Allen, 1970; Rose et al., 1976; Bridgland, 1988a; Whiteman & Rose, 1992; Whiteman, 1992). Recent re-evaluation, by the last named author, has shown that the Formation can be subdivided altitudinally into a series of member-status units comparable to those described here. The deposits identified as deposits of north-eastward flowing tributaries, the Epping Forest Formation and the Warley Gravel, should conceivably have mainstream equivalents but it is not yet possible to correlate them with specific units within the Kesgrave Formation. However, the altitude of the tributary spreads suggests that they are of considerable antiquity (cf. Gibbard, 1979; section 9.3). Extension of the assumed gradient of 1.4 m km^{-1} (section 2.1) downstream to the area of the 'Mid-Essex Depression' (c. 21–22 km), thought to be the final course of the Thames before its diversion, indicates the following upper surface heights for each Epping Forest member:

High Beach Gravel	100 m OD
Debden Green Gravel	85 m OD
Buckhurst Hill Gravel	65.5 m OD
Woodford Green Gravel	54 m OD

Table 9. *Correlation of the Lower Thames sequence with those in neighbouring areas*

See text for explanations. [1]Gibbard (1985; 1989); [2]Bridgland (1988) and Bridgland et al. (i1993); [3]Mitchell et al. (1973)

Middle Thames[1] Staines Alluvial Deposits	Lower Thames Tilbury Deposits	Southeastern Essex[2] coastal deposits	Central Essex	Stage[3] Flandrian
Shepperton Gravel Langley Silt Complex Kempton Park Gravel Reading Town Gravel	Shepperton Gravel Langley Silt Complex East Tilbury Marshes Gravel West Thurrock Gravel	unnamed ? unnamed unnamed	— — — —	late Middle Devensian early
Trafalgar Square/ Brentford deposits	Aveley, Crayford Silts and Sands	?	—	Ipswichian
Spring Gardens Gravel Taplow Gravel Lynch Hill Gravel Boyn Hill Gravel	Spring Gardens Gravel Mucking Gravel Corbetts Tey Gravel Orsett Heath Gravel	? unnamed Barling Gravel Southchurch–Asheldham Gravel (part)	— — — —	late Wolstonian early
—	Middle Gravel Swanscombe Lower Loam Lower Gravel	Southeast Essex Channel Deposits	—	Hoxnian
Black Park Gravel	Dartford Heath Gravel —initiation of Lower Thames—	Southchurch–Asheldham Gravel (part)	late —	Anglian
Middle Thames Gravel Formation	Epping Forest Formation, Worley Gravel and Darenth Wood Gravel	High-level East Essex Gravel Formation	Kesgrave Formation	early to pre-Anglian

On the basis of the stratigraphical relationships described above (section 2.1), the Woodford Green Member may be analogous to the Dollis Hill Gravel in the Mole–Wey valley or the Westmill Gravel in the Thames system, in that it could be of early Anglian age, i.e. immediately predating and contemporary with the glaciation of the region (cf. Gibbard, 1977, 1979). If so, then the higher deposits are almost certainly pre-Anglian, also by analogy with the Thames sequence (Gibbard, 1985, 1989; Whiteman & Rose, 1992). The regional gradients drawn by the last mentioned authors suggest that the High Beach Gravel may be equivalent to the Westland Green Member, the Debden Green Gravel to the Satwell Member, and the Buckhurst Hill Gravel to the Gerrards Cross Member. Detailed investigation of these and other southern tributary units would be valuable for the reconstruction of the pre-glacial Thames' drainage system.

9.2 Middle Thames Valley

For the purposes of this discussion, the Middle Thames Valley extends immediately upstream of the Lower Thames, from the arbitrary line in central London, as far as Goring. The Pleistocene sequence has been the subject of several publications by the author (Gibbard, 1983, 1985, 1989), and by Green & McGregor (1978, 1980) and Green, McGregor & Evans (1982). It includes a series of aggradations that extend back into the early Pleistocene. Up to and including the Winter Hill Gravel Member (Middle Thames Gravel Formation), the fluvial aggradations deviate from the course through London and are instead aligned to the north east via the Vale of St Albans in Hertfordshire. Glacial diversion of the Thames and several south-bank tributaries during the maximum ice advance in the Anglian Stage caused the river to adopt its course through London and initiated the present Lower Thames course (Gibbard, 1974, 1977, 1979, 1988b).

The earliest member that can be correlated with the Lower Thames sequence is therefore the first through the valley, i.e. the Black Park Gravel. The correlation of this unit with the Dartford Heath Gravel has already been discussed in detail. The Swanscombe sequence has no upstream counterpart, but the Orsett Heath Gravel passes upstream into the Boyn Hill Member. Likewise, the Corbets Tey is the continuation of the Lynch Hill and the Mucking Gravel is the continuation of the Taplow Member. The relationship of the Ipswichian Brentford Sands with the Lower Thames' Aveley/Crayford Silt and Sands Complex was noted in the previous chapter.

Of particular interest is the subsequent, early Devensian West Thurrock Gravel. In the Middle Thames a similarly placed member, the Reading Town Gravel, was identified in the Reading area by Gibbard (1985). This unit was equated with the upper part of the Summertown–Radley Terrace Gravel of the Upper Thames Valley, where there is ample evidence of its early Devensian age. However, the Reading Town Member could not be traced downstream. The identification of a unit in the same relative stratigraphical position in the Lower Thames Valley suggests that the two units may be equivalents and, if so, they demonstrate that an early Devensian aggradation did originally occur throughout the valley system. The reason for its extensive later removal is uncertain.

The following Middle Devensian members, the East Tilbury Marshes Gravel and the Kempton Park Gravel, are undoubted equivalents, and the Middle Thames' Shepperton Gravel can be followed continuously below the modern floodplain Staines/Tilbury Deposits throughout the valley.

9.3 South-eastern Essex

South-eastern Essex represents the east and north-eastward extension of the Lower Thames area. Detailed studies on the fluvial aggradations by Bridgland (1980, 1983a, 1983b, 1988a) and substantial channel fills by Roe (1994) have shown that this area contains a series of accumulations that probably began forming in the early Pleistocene. Bridgland has subdivided the sequence into two formations. The older or High-level East Essex Gravel Formation predates the arrival of the Thames and was instead deposited by a north eastward flowing River Medway, this river being confluent with the early Thames further to the north. Individual members of this formation can be shown to be equivalent to members within the Kesgrave Formation ('Group'), the latter laid down by the ancestoral Thames before its diversion (Rose & Allen, 1970; Rose et al., 1976; Bridgland, 1988a; Whiteman & Rose, 1992; Whiteman, 1992).

As discussed in section 8.2, it is thought probable that before the glacial diversion of the Thames, the Darent–Cray stream must have flowed eastwards, possibly along the approximate line of the present Thames course, to a confluence with the Medway (cf. Bridgland, 1988a). This is based on the altitude of the Darenth Wood Gravel. This spread occurs at a height at which the stream would have been prevented from continuing north of the present valley since its course would have been blocked by the bedrock Brentwood ridge. It is not possible to determine to which of the High-level East Essex Formation members the Darenth Wood Gravel is equivalent, since the gradient of this unit is unknown.

The change to the Low-level East Essex Gravel Formation is marked by an influx of local materials and exotic lithologies, the same as those in the Lower Thames Dartford Heath and subsequent members. This change clearly therefore reflects the establishment of the diverted Thames along the present course and then into the Medway Valley, the confluence occurring near Southend.

In spite of this marked change, Bridgland (1983a, b, 1988a) did not identify an equivalent of the earliest Thames unit, the Dartford Heath Gravel. The highest unit he finds is his Southchurch/Asheldham Gravel that he considered to be equivalent of the Orsett Heath Member. This he explained as possibly resulting from its 'steeper downstream gradient...which suggests [the former] may fall below the level of the Boyn Hill [Orsett Heath] aggradation between London and Southend.' However, extension of the gradient of 50 cm km^{-1} from the Orsett Cock downstream to Southminster, a distance of 39 km, gives a surface height of 19 m OD and a base at c. 14 m OD. This would bring the Dartford Heath Gravel to an elevation virtually the same as that for the Orsett Heath-equivalent Southchurch/Asheldham Gravel, contrary to Bridgland's conclusion. It would seem highly probable, therefore, that the Southchurch/Asheldham Member could be internally composite and could potentially include abutting Dartford Heath and Orsett Heath-equivalent elements. This has also been suggested by Bridgland et al. (1993).

Observations by the author in the same Southchurch/Asheldham Gravel spread, 5 km north-east at Tillingham (TL 986037: 16 m) in 1977, showed over 3–4 m of stratified gravel and sand disposed into large-scale foreset beds prograding towards the north-east. They unconformably overlaid horizontally stratified silty sands, 1–3 m thick. The sedimentary facies of these sediments was quite unlike that seen in normal gravel bed stream sediments and was almost certainly of proximal foreset and bottom set deltaic origin. If the deposits are deltaic, this may explain the lack of equivalent spreads downstream. It is conceivable that these deposits could have accumulated in a substantial water body with relatively elevated water level. This water body may have been the southern North Sea, or more likely the substantial ice-dammed lake thought to have occupied the basin during the Anglian Stage (cf. Gibbard, 1988b). Further investigations of this problem will be presented elsewhere.

The downcutting that intervened between deposition of the Dartford Heath Gravel and the Swanscombe members (Chapter 8) seems also to be represented by downcutting of a similar order in eastern Essex. Here a complex of deep channels occur, infilled predominantly by fine sediments of temperate character. According to Roe (1994), they show a transition from initially fresh to brackish water sedimentation and represent the evolution of the Thames Estuary during the Hoxnian Stage.

As already mentioned, the Orsett Heath Gravel-equivalent part of the Southchurch/Asheldham Member is the first Thames gravel unit recognised by Bridgland (1988a). However, reinterpretation of the northern part of this spread as the Dartford Heath equivalent implies that the river flowed eastwards south of Tillingham. The Orsett Heath equivalent of the Southchurch/Asheldham Member and subsequent units rest on a dissected surface in which the channel infills occur, such that the channels underlie several of the younger gravel members. Contrary to Bridgland's (1988a) interpretation, there are no grounds for thinking that they represent more than one temperate event (see below; Roe, 1994).

Downstream extension of the Corbets Tey Gravel from the Stanford-le-Hope area to the Southend area indicates that it is equivalent to the Barling Gravel of Bridgland (1988a). However, the Barling Member was correlated with the Mucking Gravel, according to this author. More recently he has retracted this correlation in favour of that proposed here (Bridgland et al., 1993). The result of this is that the Rochford Gravel of the Southend area appears to have no upstream equivalent. It is now considered by Bridgland et al. (1993) to be a dissected remnant of the Southchurch/Asheldham Member.

This re-evaluation of Bridgland's (1983a, b, 1988a) correlations with the Lower Thames Valley also affects the Mucking Gravel, which was previously thought to be equivalent to the Barling Member. It now appears that the Mucking Gravel would be below sea level in the eastern Essex area, and therefore may be represented by gravel 'below the coastal marshes' (Bridgland et al., 1993). Similarly, the younger units are also present below sea level in this area and therefore do not have equivalents on land. Nevertheless, offshore equivalents of three units (?Mucking, East Tilbury Marshes and Shepperton gravels) have been detected on the sea floor, trending parallel to the present coast as far as the Blackwater Estuary from where they turn westwards (Bridgland & D'Olier, 1989; Bridgland et al., 1993).

9 CORRELATION WITH NEIGHBOURING AREAS

It is uncertain at present whether or not equivalent sediments of the Ipswichian estuarine sequences are present in south-eastern Essex. It is possible that the Shoeburyness and Burnham channel fills are downstream representatives of this aggradation, but the local stratigraphy suggests that this is not correct and that they represent part of the earlier estuarine phase (Roe, 1994).

Therefore, as Bridgland (1980, 1983a, b, 1988a) has repeatedly emphasised, it is possible to relate the aggradations of the Lower Thames Valley to directly equivalent units in eastern Essex and even into the offshore region.

References

Aario, L. (1940). Waldgrenzen und subrezenten pollen-spektren in Petsamo, Lappland. *Ann. Acad. Sci. Fennicae* **A54**, 1–120.

Abbott, W.J.L. (1890). Notes on some Pleistocene sections in and near London. *Proc. Geol. Ass.* **11**, 473–80.

Allen, J.R.L. (1970). Studies in fluviatile sedimentation: a comparison of fining-upward cyclothems, with special reference to coarse-member composition and interpretation. *J. Sedim. Petrol.* **40**, 298–323.

Allen., P., Cheshire, D.A. & Whiteman, C.A. (1991). Events on the southern margin of the Anglian ice sheet. In: Ehlers, J., Gibbard, P.L. & Rose, J. (eds) *Glacial deposits of Britain and Ireland*, pp. 255–78. Balkema: Rotterdam.

Allison, J., Godwin, H. & Warren, S.H. (1952). Late-glacial deposits at Nazing in the Lea Valley, North London. *Phil. Trans. R. Soc. Lond.* **B236**, 169–240.

American Society for Testing and Materials (ASTM) (1964). *Procedures for testing soils*. Am. Soc. for Testing and Materials: Philadelphia, 95–101.

Andrew, R. (1970). The Cambridge pollen reference collection. In: Walker, D. & West, R.G. (eds) *Studies in the vegetational history of the British Isles*. Cambridge University Press: Cambridge, 225–231.

Baker, C.A. & Jones, D.K.C. (1980). Glaciation of the London Basin and its influence on the drainage pattern: a review and appraisal. In: Jones, D.K.C. (ed.) *The shaping of southern England*. Inst. Br. Geogr. Spec. Publ., **11**, 131–76. Academic Press: London.

Baker, C.A. (1983). Glaciation and Thames diversion in the Mid-Essex depression. In: Rose, J. (ed.) *Diversion of the Thames*, pp. 39–49. Field guide, Quaternary Research Association: Cambridge.

Barton, N. (1962). *The lost rivers of London*. Phoenix House & Leicester University Press.

Bates, C.D., Coxon, P. & Gibbard, P.L. (1978). A new method for the preparation of clay-rich sediment samples for palynological investigation. *New Phytol.* **81**, 459–63.

Berry, F.G. (1979). Late Quaternary scour hollows and related features in central London. *Q. J. Engng. Geol.* **12**, 9–29.

Birks, H.J.B. (1973). *Past and present vegetation of the Isle of Skye – a palaeoecological study*. Cambridge University Press: Cambridge.

Blackith, R.E. & Reyment, R.A. (1971). *Multivariate morphometrics*. Academic Press: London.

Blandford, W.T. (1854). On a section lately exposed in some excavations at the

West India Docks. *Q. J. Geol. Soc. Lond.* **10**, 433–5.

Blezard, R.G. (1966). Field meeting at Aveley and West Thurrock. *Proc. Geol. Ass.* **77**, 273–6.

Bluck, B.J. (1967). Sedimentation of beach gravels: an example from south Wales. *J. Sedim. Petrol.* **37**, 128–56.

Bluck, B.J. (1979). The structure of coarse grained braided stream alluvium. *Trans. R. Soc. Edinburgh* **70**, 181–221.

Boulger, G.S. (1876). Excavation in Greenwich. *Proc. W. Lond. Sci. Assoc.* **1**, 47.

Bowen, D.Q., Hughes, S., Sykes, G.A. & Miller, G.H. (1989). Land–sea correlations in the Pleistocene based on isoleucine epimerisation in non-marine molluscs. *Nature* **340**, 49–51.

Bowen, D.Q., Rose. J., McCabe, A.M. & Sutherland, D.G. (1986). Correlation of Quaternary glaciations in England, Ireland, Scotland and Wales. *Quat. Sci. Rev.* **5**, 299–340.

Bowen, D.Q. & Sykes, G.A. (1988). Correlation of marine events and glaciations on the northeast Atlantic margin. *Phil. Trans. R. Soc. Lond.* **B318**, 619–35.

Bridgland, D.R. (1980). A reappraisal of Pleistocene stratigraphy in north Kent and eastern Essex, and new evidence concerning former courses of the Thames and Medway. *Quat. Newsl.* **32**, 15–24.

Bridgland, D.R. (1983a). *The Quaternary fluvial deposits of north Kent and eastern Essex*. Ph.D. thesis, CNAA City of London Polytechnic.

Bridgland, D.R. (1983b). Eastern Essex. In: Rose, J. (ed.) *Diversion of the Thames*, pp. 170–84. Field guide, Quaternary Research Association: Cambridge.

Bridgland, D.R. (1986). The rudaceous components of the East Essex gravel; their characteristics and provenance. *Quat. Studies* **2**, 34–44.

Bridgland, D.R. (1988a). The Pleistocene fluvial stratigraphy and palaeogeography of Essex. *Proc. Geol. Ass.* **99**, 291–314.

Bridgland, D.R. (1988b). Problems in the application of lithostratigraphic classification to Pleistocene terrace deposits. *Quat. Newsl.* **55**, 1–8.

Bridgland, D.R. & D'Olier, B. (1989). A preliminary correlation of the onshore and offshore courses of the Rivers Thames and Medway during the Middle and Upper Pleistocene. In: Henriet, J.P. & de Moor, G. (eds) *The Quaternary and Tertiary geology of the Southern Bight, North Sea*, pp. 161–72. Ministry of Economic Affairs, Belgian Geological Survey.

Bridgland, D.R., D'Olier, B., Gibbard, P.L. & Roe, H.M. (1993). Correlation of Thames terrace deposits between the Lower Thames, eastern Essex and the submerged offshore continuation of the Thames–Medway valley. *Proc. Geol. Ass.*

Bridgland, D.R., Gibbard, P.L., Harding, P., Kemp, R.A. & Southgate, G. (1985). New information and results from recent excavations at Barnfield Pit, Swanscombe. *Quat. Newsl.* **46**, 25–38.

Briggs, D.J. & Gilbertson, D.D. (1980). Quaternary processes and environments in the Upper Thames Valley. *Trans. Inst. Br. Geogr.* **5**, 53–65.

Bromehead, C.E.N. (1912). On the diversions of the Bourne, near Chertsey. Summary of progress for (1911. *Geol. Surv. G.B.* Appendix II.

Bromehead, C.E.N. (1925). The geology of North London. *Mem. Geol. Surv. G.B.*

Bryant, I.D. (1983a). The utilisation of arctic analogue studies in the interpretation of periglacial river sediments from southern Britain. In: Gregory, K.J. (ed.) *Background to palaeohydrology*, pp. 415–31. J.Wiley & Sons: Chichester.

Bryant, I.D. (1983b). Facies sequences associated with some braided river deposits of Late Pleistocene age from southern Britain. In: Collinson, J.D. (ed.) *Modern and ancient fluvial systems*, pp. 267–75. Proc. Int. Fluvial Conf. Keele 1982.

Burchell, J.P.T. (1931). Early Neanthropic man and his relation to the Ice Age. *Proc. Prehist. Soc. East Anglia* **6**, 253–303.

Burchell, J.P.T. (1933). The Northfleet 50-foot submergence later than the Coombe rock of post-Early Mousterian times. *Archaeologia* **83**, 67–92.

Burchell, J.P.T. (1934a). The Middle Mousterian Culture and its relation to the Coombe Rock of post-early Mousterian times. *Antiquaries Journal* **14**, 33–9.

Burchell, J.P.T. (1934b). Fresh facts relating to the Boyn Hill Terrace of the Lower Thames Valley. *Antiquaries Journal* **14**, 163–6.

Burchell, J.P.T. (1935). Some Pleistocene deposits at Kirmington and Crayford. *Geol. Mag.* **72**, 327–31.

Burchell, J.P.T. (1936a). Evidence of a Late Glacial episode within the valley of the Lower Thames. *Geol. Mag.* **73**, 91–2.

Burchell, J.P.T. (1936b). A final note on the Ebbsfleet channel series. *Geol. Mag.* **73**, 550–4.

Burchell, J.P.T. (1954). Loessic deposits in the Fifty-foot terrace post-dating the Main Coombe Rock of Baker's Hole, Northfleet, Kent. *Proc. Geol. Ass.* **65**, 256–61.

Burchell, J.P.T. (1957). A temperate bed of the Last Interglacial at Northfleet, Kent. *Geol. Mag.* **94**, 212–14.

Callow, P. (1976). *The Lower and Middle Palaeolithic of Britain and adjacent areas*. Ph.D. thesis, University of Cambridge.

Cant, D.J. (1976). Selective preservation of flood stage deposits in a braided fluvial environment. *Geol. Soc. Ca. Progr. Abstr.* **1**, 77.

Carreck, J. (1976). Pleistocene mammalia and molluscan remains from 'Taplow' terrace deposits at West Thurrock, near Grays, Essex. *Proc. Geol. Ass.* **87**, 83–91.

Chandler, R.H. (1914). The Pleistocene deposits at Crayford. *Proc. Geol. Ass.* **25**, 61–70.

Chandler, R.H. (1916). The implements and cores of Crayford. *Proc. Prehist. Soc. East Anglia* **2**, 240–8.

Chandler, R.H. & Leach, A.L. (1908). Excursion to Crayford and Dartford Heath. *Proc. Geol. Ass.* **20**, 122–6.

Chandler, R.H. & Leach, A.L. (1912a). On the Dartford Heath Gravel and on a Palaeolithic implement factory. *Proc. Geol. Ass.* **23**, 102–111.

Chandler, R.H. & Leach, A.L. (1912b). Report of an excursion to the Lower Tertiary section and the Pleistocene river drifts near Erith. *Proc. Geol. Ass.* **23**, 183–90.

Cheshire, D.A. (1981). A contribution towards a glacial stratigraphy of the Lower Lea Valley and implications for the Anglian Thames. *Quat. Studies*, **1**, 27–69.

Cheshire, D.A. (1983). Till lithology in Hertfordshire and west Essex. In: Rose, J. (ed.) *Diversion of the Thames*, pp. 50–9. Field guide, Quaternary Research Association: Cambridge.

Cheshire, D.A. (1986). *The lithology and stratigraphy of the Anglian deposits of the Lea basin*. Ph.D. thesis. CNAA Hatfield Polytechnic.

Churchill, D.M. (1965). The displacement of deposits at sea level, 6500 years ago in southern Britain. *Quaternaria* **7**, 239–49.

Clayton, K.M. (1977). River terraces. In: Shotton, F.W. (ed.) *British Quaternary studies*, pp. 153–67. Oxford University Press: Oxford.

REFERENCES

Collinson, J.D. (1970). Bed forms in the Tana River, Norway. *Geogr. Ann.* **52A**, 31–55.

Conway, B.W. (1971). Geological investigations of Boyn Hill Terrace deposits at Barnfield Pit, Swanscombe, Kent during (1970. *Proc. R. Anthropol. Inst.*, **1971**, 60–64.

Conway, B.W. (1973). Geological investigations of Boyn Hill Terrace deposits at Barnfield Pit, Swanscombe, Kent during 1971. *Proc. R. Anthropol. Inst.*, **1973**, 80–5.

Coope, G.R. (1974). Interglacial Coleoptera from Bobbitshole, Ipswich, Suffolk. *J. Geol. Soc. Lond.* **130**, 333–40.

Cooper, J. (1972). Last interglacial (Ipswichian) non-marine Mollusca from Aveley, Essex. *Essex Naturalist* **33**, 9–14.

Corner, F. (1903). Palaeolithic implements from Leyton. *Essex Naturalist*, **13**, 84.

Corner, R. (1975). The Tana Valley terraces. A study of the morphology and sedimentology of the lower Tana Valley terraces between Mansholmen and Maskejokka, Finmark, Norway. *Uppsala Universitet Naturgeografiska Institutionen Rapport* **38**.

Cotton, C.A. (1940). Classification and correlation of river terraces. *J. Geomorph.* **3**, 27–37.

Cotton, R.P. (1847). On the Pliocene deposits of the valley of the Thames at Ilford. *Ann. Nat. Hist.* **20**, 164–9.

Davies, W., Woodward, H. & Brady, A. (1874). *Catalogue of the Pleistocene Vertebrata from the neighbourhood of Ilford, Essex, etc.* (private circulation).

Davis, R.A. (1985). *Coastal sedimentary environments.* Springer-Verlag: New York.

Dawkins, W.B. (1867). On the age of the Lower Brickearths of the Thames Valley. *Q. J. Geol. Soc. Lond.* **23**, 91–109.

Dawson, M.R. (1985). Environmental reconstructions of a late Devensian terrace sequence. Some preliminary findings. *Earth Surf. Proc. Landf.* **10**, 237–46.

Dawson, M.R. & Bryant, I.D. (1987). Three-dimensional facies geometry in Pleistocene outwash sediments, Worcestershire, UK. In: Ethridge, F.G., Flores, R.M. & Harvey, M.D. (eds) *Recent developments in fluvial sedimentology.* Soc. Econ. Paleont. Minerol. Special Publ. **39**, 191–6.

Devoy, R.J.N. (1979). Flandrian sea level changes and vegetational history of the lower Thames Valley. *Phil. Trans. R. Soc. Lond.* **B 285**, 355–410.

Dewey, H. (1926). Palaeoliths in Hyde Park. *Antiq. J.* **6**, 73–5.

Dewey, H. (1932). The Palaeolithic deposits of the Lower Thames Valley. *Q. J. Geol. Soc. Lond.* **88**, 35–54.

Dewey, H. & Bromehead, C.E.N. (1915). The geology of the country around Windsor and Chertsey. *Mem. geol. Surv. G.B.*

Dewey, H. & Bromehead, C.E.N. (1921). The geology of south London. *Mem. Geol. Surv. G.B.*

Dewey, H., Bromehead, C.E.N., Chatwin, C.P & Dines, H.G. (1924). The geology of the country around Dartford. *Mem. Geol. Surv. G.B.*

Dines, H.G. & Edmunds, F.H. (1925). The geology of the country around Romford. *Mem. Geol. Surv. G.B.*

Evans, C. (1863). On a superficial freshwater deposit near Blackfriars Road. *Proc. Geol. Ass.* **1**, 264–7.

Evans, J. (1860). On the occurrence of flint implements in undisturbed beds of gravel, sand and clay. *Archaeologia* **38**, 280–307.

REFERENCES

Evans, P. (1971). Towards a Pleistocene time scale. Pt II, The Phanerozoic time scale – a supplement. *Special Publication of the Geological Society of London* **5**, 123–356.

Fisk, H.N. (1944). *Geological investigations in the alluvial valley of the lower Mississippi River*. Corps of Engineers, U.S. Army.

Franks, J.W. (1960). Interglacial deposits at Trafalgar Square, London. *New Phytol.* **59**, 145–52.

Friend, P.F. & Moody-Stuart, M. (1972). Sedimentation of the Wood bay formation (Devonian) of Spitsbergen: regional analysis of a late orogenic basin. *Norsk Polarint. Skr.* **15**, 1–77.

Galloway, R.W. (1961). Solifluction in Scotland. *Scott. Geogr. Mag.* **77**, 75–87.

Gibbard, P.L. (1974). *Pleistocene stratigraphy and vegetational history of Hertfordshire*. Ph.D. thesis, University of Cambridge.

Gibbard, P.L. (1977). Pleistocene history of the Vale of St. Albans. *Phil. Trans. R. Soc. Lond.* **B 280**, 445–83.

Gibbard, P.L. (1979). Middle Pleistocene drainage in the Thames Valley. *Geol. Mag.* **116**, 35–44.

Gibbard, P.L. (1982). Terrace stratigraphy and drainage history of the Plateau Gravels of north Surrey, south Berkshire and north Hampshire. *Proc. Geol. Ass.* **93**, 369–84.

Gibbard, P.L. (1983). The diversion of the Thames – a review. In: Rose, J. (ed.) *Diversion of the Thames*, pp. 8–23, Field guide, Quaternary Research Association: Cambridge.

Gibbard, P.L. (1985). *Pleistocene history of the Middle Thames Valley*. Cambridge University Press: Cambridge.

Gibbard, P.L. (1988a). Problems in the application of lithostratigraphic classification to Pleistocene terrace deposits. *Quat. Newsl.* **56**, 22–3.

Gibbard, P.L. (1988b). The history of the great northwest European rivers during the past three million years. *Phil. Trans. R. Soc. Lond.* **B 318**, 559–602.

Gibbard, P.L. (1989). The geomorphology of a part of the middle Thames forty years on: a reappraisal of the work of F. Kenneth Hare. *Proc. Geol. Ass.* **100**, 481–503.

Gibbard, P.L. & Stuart, A.J. (1975). Flora and vertebrate fauna of the Barrington Beds. *Geol. Mag.* **112**, 493–501.

Gibbard, P.L. & Turner, C. (1990). Cold stage type sections: some thoughts on a difficult problem. *Quaternaire* **1**, 33–40.

Gibbard, P.L., Whiteman, C.A. & Bridgland, D.R. (1988). A preliminary report on the stratigraphy of the Lower Thames Valley. *Quat. Newsl.* **56**, 1–8.

Gibbard, P.L., Wintle, A.G. & Catt, J.A. (1987). The age and origin of clayey silt brickearth, West London, England. *J. Quatern. Sci.* **2**, 3–9.

Godwin, H. & Willis E.H. (1962). Cambridge University natural radiocarbon measurements V. *Radiocarbon* **4**, 57–70.

Gordon, A.D. & Birks, H.J.B. (1974). Numerical methods in palaeoecology: II. Comparison of pollen diagrams. *New Phytol.* **73**, 221–49.

Green, C.P. & McGregor, D.F.M. (1978). Pleistocene gravel trains of the River Thames. *Proc. Geol. Ass.* **89**, 143–56.

Green, C.P. & McGregor, D.F.M. (1980). The terraces of the River Thames and their palaeohydrological implications. In: Jones, D.K.C. (ed.) *The shaping of southern England*. Inst. Br. Geogr. Spec. Publ., **11**, 131–76. Academic Press: London.

REFERENCES

Green, C.P., McGregor, D.F.M. & Evans, A. (1982). Development of the Thames drainage system in Buckinghamshire and Hertfordshire. *Geol. Mag.* **119**, 281–90.

Greenhill, J.E. (1884). The implementiferous gravels of north-east London. *Proc. Geol. Ass.* **8**, 336–43.

Hanson, C.B. (1980). Fluvial taphonomic processes: models and experiments. In: Behrensmeyer, A.K. & Hill, A.P. (eds) *Fossils in the making*, pp. 156–81. University of Chicago Press: Chicago.

Harding, P. & Gibbard, P.L. (1984). Excavations at Northwold Road, Stoke Newington, north east London, 1981. *Trans. Lond. Middx. Archaeol. Soc.* **34**, 1–18.

Harding, P., Gibbard, P.L., Lewin, J., Macklin, M.G. & Moss, E.H. (1985). The transport and abrasion of flint hand-axes in a gravel-bed river. In: Sieveking, G. & Newcomer, M. (eds) *The human uses of flint and chert: proceedings of the fourth international Flint symposium*, pp. 115–26. Cambridge University Press: Cambridge.

Hare, F.K. (1947). The geomorphology of part of the Middle Thames. *Proc. Geol. Ass.* **58**, 294–339.

Harms, J.C. & Fahnestock, R.K. (1965). Stratification, bedforms and flow phenomena (with an example from the Rio Grande). In: Middleton, G.V. (ed.) *Primary sedimentary structures and their hydrodynamic interpretation. Soc. Econ. Palaeontol. Mineral. Spec. Publ.* **12**, 84–115.

Harms, J.C., Southard, J.B., Spearing, D.R. & Walker, R.G. (1975). depositional environments as interpreted from primary sedimentary structures and stratification sequences. *Soc. Econ. Palaeontol. Mineral.* Short Course 2.

Hayward, J.F. (1955). Borehole records from the Lea Valley between Cheshunt and Edmonton. *Proc. Geol. Ass.* **66**, 68–76.

Hayward, J.F. (1956). Certain abandoned channels of Pleistocene and Holocene age in the Lea Valley, and their deposits. *Proc. Geol. Ass.* **69**, 32–63.

Hedberg, H.D. (1976). *International stratigraphic guide*. J. Wiley & Sons: New York.

Hein, F.J. & Walker, R.G. (1977). Bar evolution and development of stratification in the gravelly, braided, Kicking Horse River, B.C. *Can. J. Earth Sci.* **14**, 562–70.

Hey, R.W. (1965). Highly quartzose pebble gravels in the London Basin. *Proc. Geol. Ass.* **76**, 403–20.

Hey, R.W. (1976). The terraces of the Middle and Lower Thames. *Studia Soc. Sci. Torunensis.* **8C**, 115–22.

Hinton, M.A.C. (1900). The Pleistocene deposits of the Ilford and Wanstead district. *Proc. Geol. Ass.* **16**, 271–7.

Hinton, M.A.C. & Kennard, A.S. (1901). Contributions to the Pleistocene geology of the Thames Valley, 1: The Grays Thurrock area, pt. 1. *Essex Naturalist* **11**, 336–70.

Hinton, M.A.C. & Kennard, A.S. (1905). The relative ages of the stone implements of the Lower Thames Valley. *Proc. Geol. Ass.* **19**, 76–100.

Hinton, M.A.C. & Kennard, A.S. (1907). Contributions to the Pleistocene geology of the Thames Valley, 1: The Grays Thurrock area, pt. 2. *Essex Naturalist* **15**, 56–88.

Hinton, M.A.C. & Kennard, A.S. (1910). Excursion to Grays Thurrock, Essex. *Proc. Geol. Ass.* **21**, 474–6.

Hollin, J. (1977). Thames interglacial sites, Ipswichian sea levels and Antarctic ice surges. *Boreas* **6**, 33–52.

Holmes, T.V. (1892). The new railway from Grays Thurrock to Romford: sections between Upminster and Romford. *Q. J. Geol. Soc. Lond.* **48**, 365–72.

Holmes, T.V. (1894). Further notes on the new railway from Romford to Upminster and on the relations of the Thames Valley beds to the boulder clay. *Q. J. Geol. Soc. Lond.* **50**, 443–52.

Holmes, T.V. (1901). Geological notes on the new reservoirs in the valley of the Lea near Walthamstow, Essex. *Essex Naturalist* **12**, 1.

Holmes, T.V. (1902). Additional notes on the sections shown at the new reservoirs in the valley of the Lea near Walthamstow. *Essex Naturalist* **12**, 224.

Holyoak, D.T. (1983). A Late Pleistocene interglacial flora and molluscan fauna from Thatcham, Berkshire, with notes on Mollusca from the interglacial deposits at Aveley, Essex. *Geol. Mag.* **120**, 623–9.

Howard, A.D. (1959). Numerical systems of terrace nomenclature: a critique. *J. Geol.* **67**, 239–43.

Howard, A.D., Fairbridge, R.W. & Quinn, J.H. (1968). Terraces – fluvial. In: Fairbridge, R.W. (ed.) *The encyclopedia of geomorphology*. Reinhold Book Corp.: New York, 1117–23.

Hutchinson, J.N. (1991). Theme lecture: periglacial and slope processes. In: Forster, A., Culshaw, M.G., Cripps, J.C., Little, J.A. & Moon, C.F. (eds) *Quaternary engineering geology*. Geol. Soc. Engng. Spec. Publ. **7**, 283–331.

Jessen, K. & Milthers, V. (1928). Stratigraphical and palaeontological studies of interglacial freshwater deposits in Jutland and northwest Germany. *Danmarks Geol. Unders.* RIII, 48.

Johnson J.P. (1902). Palaeolithic implements from the low-level drift of the Thames Valley. *Essex Naturalist* **12**, 52–7.

Jones, D.K.C. (1981). *Southeast and southern England*. Methuen: London.

Jouanneau, J.M. & Latouche, C. (1981). *The Gironde Estuary*. Contrbutions to sedimentology **10**. E. Schweizerbart'sche Verlagsbuchhandlung: Stuttgart.

Kemp, R.A. (1985). The decalcified Lower Loam at Swanscombe, Kent: a buried Quaternary palaeosol. *Proc. Geol. Ass.* **96**, 343–55.

Kemp, R.A. (1991). Micromorphology of the buried Quaternary soil within Burchell's 'Ebbsfleet Channel', Kent. *Proc. Geol. Ass.* **102**, 275–88.

Kennard, A.S. (1904). Notes on a palaeolith from Grays, Essex. *Essex Naturalist* **13**, 112–13.

Kennard, A.S. (1916). The Pleistocene succession in England. *Proc. Prehist. Soc. E. Anglia* **2**, 249–67.

Kennard, A.S. (1942a). Pleistocene chronology (discussion). *Proc. Geol. Ass.* **53**, 24–5.

Kennard, A.S. (1942b). Faunas from the high terrace at Swanscombe. *Proc. Geol. Ass.* **53**, 105.

Kennard, A.S. (1944). The Crayford brickearths. *Proc. Geol. Ass.* **55**, 121–63.

Kennard, A.S. & Woodward, H.B. (1897). The post-Pliocene non-marine Mollusca of Essex. *Essex Naturalist* **10**, 87–103.

Kennard, A.S & Woodward, H.B. (1900). The Pleistocene non-marine Mollusca of Ilford. *Proc. Geol. Ass.* **16**, 282–6.

Kennard, A.S. & Woodward, H.B. (1902). The post-Pliocene non-marine Mollusca of the south of England. *Proc. Geol. Ass.* **17**, 213–60.

Kerney, M.P. (1971). Interglacial deposits in Barnfield pit, Swanscombe and their molluscan faunas. *J. Geol. Soc. Lond.* **127**, 69–93.

Kerney, M.P. & Sieveking, G. de G. (1977). Northfleet. In: Shephard-Thorn, E.R. & Wymer, J.J. (eds) *Southeast England and the Thames Valley: INQUA Guidebook for excursion A5.* Geobooks: Norwich, 44–9.

King, W.B.R. & Oakley, K.P. (1936). The Pleistocene succession in the lower part of the Thames Valley. *Proc. Prehist. Soc.* **2**, 52–76.

Leach, A.L. (1906). Excursion to Erith and Crayford. *Proc. Geol. Ass.* **19**, 137–41.

Leach, A.L. (1912). The geology of Shooter's Hill, Kent; the Shooter's Hill Gravel. *Proc. Geol. Ass.* **23**, 119–24.

Leach, A.L. (1913). On the buried channels in the Dartford Heath Gravel. *Proc. Geol. Ass.* **24**, 337–44.

Leach, A.L. (1920). Excavation in Castle Wood, Shooter's Hill. Mineral composition of the Shooter's Hill Gravel. Supplementary notes on the geology of Shooter's Hill, Kent. *Proc. Geol. Ass.* **31**, 127–32.

LeFloch, P. (1961). *Pragation de la maree dans l'estuarie de la Seine et en Seine maritime.* Thèse D.E., Université de Paris.

Leopold, L.B., Wolman, M.G. & Miller, J.P. (1964). *Fluvial processes in geomorphology.* W.H. Freeman: San Francisco.

Macklin, M.G. (1981). Prestwich's 'Southern Drift' in S.E. London: a re-evaluation. *Quat. Newsl.* **34**, 19–26.

Marston, A.T. (1937). The Swanscombe skull. *J. R. Anthropol. Inst.* **67**, 339–406.

Marston, A.T. (1942). Flint industries of the High Terrace at Swanscombe. *Proc. Geol. Ass.* **53**, 106.

Martinson, D.G., Pisias, N.G., Hays, J.D., Imbrie, J., Moore, T.C. jr., & Shackleton, N.J. (1987). Age dating and the orbital theory of the ice ages: development of a high-resolution 0 to 300 000 year chronostratigraphy. *Quat. Res.* **27**, 1–29.

Meijer, T. (1988). Fossiele Zoetwaternerieten mit het Nederlandse Kwartair en enkele opmerkingen over het voorkomen van deze groep in het Kwartair van Noordwest Europa. *De Kreukel* Jubilenummer, 12 November 1988, 89–108.

Miall, A.D. (1977). A review of the braided river depositional environment. *Earth Sci. Rev.* **13**, 597–604.

Miall, A.D. (1978). Lithofacies types and vertical profile models in braided river deposits: a summary. In: Miall, A.D. (ed.) *Fluvial sedimentology. Can. Soc. Petrol. Geol.* **5**, 597–604.

Middleton, G.W. (1973). Johannes Walther's law of correlation of facies. *Geol. Soc. Am. Bull.* **84**, 979–88.

Miller, G.H. & Hare, P.E. (1980). Amino-acid geochronology: integrity of the carbonate matrix and potential of molluscan fossils. In: Hare, P.E., Hoering, T.C. & King K. jr. (eds) *The biochemistry of amino-acids.* J.Wiley & Sons, New York. 413–43.

Miller, G.H., Hollin, J.T. & Andrews, J.T. (1979). Aminostratigraphy of UK Pleistocene deposits. *Nature* **281**, 539–43.

Mitchell, G.F., Penny, L.F., Shotton, F.W. & West, R.G. (1973). *A correlation of the Quaternary deposits of the British Isles. Geol. Soc. Lond. Spec. Rep.* **4**.

Monkton, H.W. (1894). On the gravels near Barkingside, Wanstead and Walthamstow, Essex. *Essex Naturalist* **7**, 115–20.

Morris., J. (1836). On a fresh-water deposit, containing mammalian remains recently discovered at Grays, Essex. *Ann. Mag. Nat. Hist.* **9**, 261–4.

Morris, J. (1838). On the deposits of the Thames containing Carnivora and other Mammalia in the valley of the Thames. *Mag. Nat. Hist.* **2**, 540.

Nichols, M.M. & Biggs, R.B. (1985). Estuaries. In: Davis, R.A.(ed.) *Coastal sedimentary environments*, pp. 77–186. Springer-Verlag: New York.

Oakley, K.P. (1952). Swanscombe Man. *Proc. Geol. Ass.* **63**, 271–300.

Ovey, C.D. (ed.) (1964). *The Swanscombe skull. A survey of research on a Pleistocene site.* R. Anthropol. Inst. Occas. Paper. No. 20.

Palmer, S. (1975). A Palaeolithic site at North Road, Purfleet, Essex. *Essex Archaeol. and Hist.* **7**, 1–13.

Parks, D.A. & Rendel, H.M. (1992). Thermoluminescence dating and geochemistry of loessic deposits in southeast England. *J. Quatern. Sci.* **7**, 99–107.

Paterson, T.T. (1940). The Swanscombe skull: a defence. *Proc. Prehist. Soc.* **6**, 166–9.

Pattison, S.R. (1863). Deptford Gravel. *Geologist* **1**, 234–5.

Peake, D.S. (1971). The age of the Wandle gravels in the vicinity of Croydon. *Proc. Croydon Nat. Hist. Scient. Soc.* **14**, 145–76.

Peake, D.S. (1982). A reappraisal of the Pleistocene history of the River Wandle and its basin. *Proc. Croydon Nat. Hist. Scient. Soc.* **17**, 89–116.

Perrin, R.M.S., Rose, J. & Davis, H. (1979). The distribution, variation and origins of pre-Devensian tills in eastern England. *Phil. Trans. R. Soc. Lond.* **B287**, 535–70.

Phillips, J. (1871). *Geology of Oxford and the valley of the Thames.* Clarendon Press: Oxford.

Phillips, L.M. (1974). Vegetational history of the Ipswichian/Eemian interglacial in Britain and continental Europe. *New Phytol.* **73**, 589–604.

Pocock, T.I. (1903). On the drifts of the Thames Valley near London. 'Summary of progress for 1902'. *Geol. Surv. G.B.* Appendix VIII.

Potter, P.E. & Pettijohn, F.J. (1963). *Paleocurrents and basin analysis.* Springer-Verlag: Berlin.

Preece, R.C. (1988). A second British interglacial record of *Margaritifera curicularia*. *J. Conchol.* **33**, 50–1.

Prestwich, J. (1855). On a fossiliferous deposit in the gravel at West Hackney. *Q. J. Geol. Soc. Lond.* **11**, 107–10.

Prestwich, J. (1857). *The ground beneath us; its geological phases and changes, etc.* J.van Voorst: London.

Prestwich, J. (1890). On the relation of the Westleton Beds or pebbly sands of Suffolk, to those of Norfolk and its extension inland, etc. *Q.J. Geol. Soc. Lond.* **46**, 120–54.

Prestwich, J. (1891). On the age, formation and successive drift stages of the valley of the Darent, etc. *Q. J. Geol. Soc. Lond.* **47**, 121–63.

Quinn, J.H. (1957). Paired river terraces and Pleistocene glaciation. *J. Geol.* **65**, 149–66.

Redman, J.B. (1845). The Terrace Pier, Gravesend, Kent. *Proc. Inst. Civil Engng.* **4**, 229.

Reid, C. (1877). On the origin of dry Chalk valleys and of the Coombe Rock. *Q. J. Geol. Soc. Lond.* **43**, 364–73.

Reid, C. (1897). Pleistocene plants from Casewick, Shacklewell and Grays. *Q. J. Geol. Soc. Lond.* **53**, 463–4.

Reineck, H.-E. & Singh, I.B. (1980). *Depositional sedimentary environments.* second edition Springer-Verlag: Berlin.

Roberts, M.B. (1986). Excavation of the Lower Palaeolithic site at Amey's Eartham Pit, Boxgrove, West Sussex; a preliminary report. *Proc. Prehist. Soc.* **52**, 215–45.

Roe, D.A. (1968a). Gazetteer of Lower and Middle Palaeolithic sites. *Council Brit. Archaeol. Res. Rep.* **8**.

Roe, D.A. (1968b). British Lower and Middle Palaeolithic hand-axe groups. *Proc. Prehist. Soc.* **34**, 1–82.

Roe, D.A. (1981). *The Lower and Middle Palaeolithic Periods in Britain*. Routledge and Kegan Paul: London.

Roe, H. (1994). *Pleistocene buried valleys in Essex*. Ph.D. thesis. University of Cambridge.

Rolfe, W.D. (1958). A recent temporary section through the Pleistocene deposits at Ilford. *Essex Nat.* **30**, 93–101.

Rose, J. & Allen, P. (1977). Middle Pleistocene stratigraphy in south-east Suffolk. *Q. J. Geol. Soc. Lond.* **133**, 83–102.

Rose, J. Allen, P. & Hey, R.W. (1976). Middle Pleistocene stratigraphy in southern East Anglia. *Nature London*, **236**, 492–4.

Roy, P.S, Thom, B.G. & Wright, L.D. (1980). Holocene sequences on an embayed coast: an evolutionary model. *Sedim. Geol.* **26**, 1–19.

Rust, B.R. (1972). Structure and process in a braided river. *Sedimentology* **18**, 221–45.

Rust, B.R. (1978a). A classification of alluvial channel systems. In: Miall, A.D. (ed.) *Fluvial sedimentology*, pp. 187–98. Canadian Soc. Petrol. Geol., Mem. 5.

Rust, B.R. (1978b). Depositional models for braided alluvium. In: Miall, A.D. (ed.) *Fluvial sedimentology*, pp. 605–25. Canadian Soc. Petrol. Geol., Mem. 5.

Rust, B.R. & Koster, E.H. (1984). Coarse alluvial deposits. In: Walker, R.G. (ed.) *Facies models*. Geoscience Canada reprint series 1, 53–69.

Salomon, J.C. & Allen, G.P. (1983). Rôle sedimentologique de la maree dans les estuaries à fort manage. Companie français des Petroles. *Notes et Mémoires* **18**, 35–44.

Salter, A.E. (1905). On the superficial deposits of central and parts of southern England. *Proc Geol. Ass.* **19**, 1–56.

Shackleton, N.J. & Opdyke, N.D. (1976). Oxygen isotope and palaeomagnetic stratigraphy of Pacific core V28-239, Late Pliocene to latest Pleistocene. *Mem. Geol. Soc. America* **145**, 449–64.

Shackley, M. (1974). Stream abrasion of flint implements. *Nature London* **248**, 501–2.

Sherlock, R.L. & Noble, A.H. (1922). The geology of the country around Beaconsfield. *Mem. Geol. Surv. G.B.*

Shotton, F.W. & Williams, R.E.G. (1971). Birmingham University radiocarbon dates V. *Radiocarbon* **13**, 141–56.

Singer, R., Wymer, J.J., Gladfelter, B.G. & Wolff, R. (1973). Excavation of the Clactonian Industry at the golf course Clacton-on-Sea, Essex. *Proc. Prehist. Soc.* **39**, 6–74.

Smith, H.J.U. (1949). Physical effects of Pleistocene climatic changes in non-glaciated areas: eolian phenomena, frost action and stream terracing. *Bull. Geol. Soc. Am.* **60**, 1485–516.

Smith, R.A. & Dewey, H. (1913). Stratification at Swanscombe. *Archaeologia* **64**, 177–204.

Smith, R.A. & Dewey, H. (1914). The high terrace of the Thames: report on excavations made on behalf of the British Museum and H.M. Geological Survey in 1913. *Archaeolgia* **65**, 187–212.

Smith, R.A. (1911). A Palaeolithic industry at Northfleet, Kent. *Archaeologia* **62**, 515–532.

Smith, W.G. (1879). On Palaeolithic implements from the valley of the Lea. *J. R. Anthropol. Inst.* **8**, 275–9.

Smith, W.G. (1884). On a Palaeolithic floor at North East London. *Trans. Essex Fld Club* **3**, 76.

Smith, W.G. (1887). On a Palaeolithic floor at North-east London. *J. R. Anthropol. Inst.* **13**, 357–84.

Smith, W.G. (1894). *Man the primaeval savage*. London.

Snelling, A.J.R. (1964). Excavations at the Globe Pit, Little Thurrock, Grays, Essex, 1961. *Essex Naturalist* **31**, 199–208.

Snelling, A.J.R. (1973/4). A fossil molluscan fauna at Purfleet, Essex. *Essex Naturalist* **33**, 104–8.

Sparks, B.W. & West, R.G. (1959). The palaeoecology of the interglacial deposits at Histon Road, Cambridge. *Eiszeit. u. Gegenw.* **10**, 123–43.

Spurrell, F.C.J. (1880). On the discovery of the place where Palaeolithic implements were made at Crayford. *Q. J. Geol. Soc. Lond.* **26**, 544–8.

Spurrell, F.C.J. (1885). Excursion to Erith and Crayford. *Proc. Geol. Ass.* **9**, 213–16.

Spurrell, F.C.J. (1886). A sketch of the history of the rivers and denudation of west Kent. *Rep. W. Kent Nat. Hist. Soc.* 1–25.

Spurrell, F.C.J. (1889). On the estuary of the River Thames and its alluvium. *Proc. Geol. Ass.* **11**, 210–30.

Stopes, M. (1903). Palaeolithic implements from Shelly gravel pit at Swanscombe, Kent. *Rep. Brit. Ass. Adv. Sci.* 803–5.

Stuart, A.J. (1974). Pleistocene history of the British vertebrate fauna. *Biol. Rev.* **49**, 225–66.

Stuart, A.J. (1976). The history of the mammal fauna during the Ipswichian/Last Interglacial in England. *Phil.Trans. R. Soc. Lond.* **B276**, 221–50.

Stuart, A.J. (1982). *Pleistocene vertebrates in the British Isles*. Longman: London.

Stuart, A.J. (1984). Pleistocene occurrence of *Hippopotamus* in Britain. *Quartärpaläontol.* **6**, 209–18.

Sutcliffe, A.J. (1964). The mammalian fauna. In: Ovey, C.D. (ed.) *The Swanscombe Skull. R. Anthropol. Inst. Occas. Pap.*, **20**, 85–111.

Sutcliffe, A.J. (1975). A hazard in the interpretation of glacial–interglacial sequences. *Quat. Newsl.* **17**, 1–3.

Sutcliffe, A.J. (1976). The British glacial – interglacial sequence. *Quatern. Newsl.* **18**, 1–7.

Sutcliffe, A.J. (1985). *On the track of ice age mammals*. British Museum: London.

Szabo, B.J. & Collins, D. (1975). Ages of fossil bones from British interglacial sites. *Nature* **25**, 680–2.

Tester, P.J. (1951). Palaeolithic flint implements from Bowman's Lodge gravel pit, Dartford Heath. *Archaeologia cantiana* **63**, 122–34.

Turner, C. (1985). Problems and pitfalls in the application of palynology to Pleistocene archaeological sites in western Europe. In: Renault-Miskowsky, J., Bui-Thi Mäi & Girard, M. Notes & monographies techniques No. 17, pp. 347–73. CNRS Centre de recherches archéol.

Turner, C. & West, R.G. (1968). The subdivision and zonation of interglacial periods. *Eiszeitu. Gegenw.* **19**, 93–101.

Tylor, A. (1868). Discovery of a Pleistocene fresh-water deposit, with shells, at Highbury New Park, near Stoke Newington. *Geol. Mag.* **5**, 391–2.

Tylor, A. (1869). On Quaternary gravels. *Q. J. Geol. Soc. Lond.* **25**, 57–99.

Waechter, J.d'A. (1969). Swanscombe (1968). *Proc. R. Anthrop. Inst.* **1968**, 53–61.

Waechter, J.d'A. (1970). Swanscombe (1969). *Proc. R. Anthrop. Inst.* **1969**, 83–93.

Waechter, J.d'A. (1971). Swanscombe (1970). *Proc. R. Anthrop. Inst.* **1971**, 43–64.

Waechter, J.d'A. (1973). Swanscombe (1971). *Proc. R. Anthrop. Inst.* **1973**, 73–85.

Waechter, J.d'A. & Conway, B.W. (1977). Barnfield Pit, Swanscombe. In: Shephard-Thorn, E.R. & Wymer, J.J. (eds.) *Southeast England and the Thames Valley*, pp. 38–55. Guide book for excursion A5. X INQUA Congress.

Walker, R.G. & Cant, D.J. (1984). Sandy fluvial systems. In: Walker, R.G. (ed.) *Facies models*. Geoscience Canada reprint series 1, 71–89.

Ward, G.R. (1984). Interglacial fossils from Upminster, Essex. *Lond. Nat.* **63**, 24–6.

Warren, S.H. (1912). A Late Glacial stage in the valley of the River Lea, subsequent to the epoch of River Drift Man. *Q. J. Geol. Soc. Lond.* **68**, 213–51.

Warren, S.H. (1916). Further observations on the Late Glacial or Ponder's End Stage of the Lea Valley. *Q. J. Geol. Soc. Lond.* **71**, 164–82.

Warren, S.H. (1918). Prehistory in Essex as recorded in the Journal of the Essex Field Club. *Essex Fld Club Spec. Mem.* **5**, 1–4.

Warren, S.H. (1923). The Late-Glacial Stage of the Lea Valley (Third report). *Q. J. Geol. Soc. Lond.* **79**, 603–5.

Warren, S.H. (1938). The correlation of the Lea Valley Arctic Beds. *Proc. Prehist. Soc.* **4**, 328–9.

Warren, S.H. (1940). Geological and prehistoric traps. *Essex Naturalist* **27**, 2–18.

Warren, S.H. (1942). The drifts of south-western Essex, Part II. *Essex Naturalist* **27**, 171–9.

Washburn, A.L. (1968). Weathering, frost action and patterned ground in the Mesters Vig district, northeast Greenland. *Medd. om Grønland* **176**.

Watson, E. (1969). The slope deposits of the Nant Iago valley near Cader Idris, Wales. *Buil. Periglac.* **18**, 95–113.

West, R.G. (1957). Interglacial deposits at Bobbitshole, Ipswich. *Phil.Trans. R. Soc. Lond.* **B241**, 1–31.

West, R.G. (1968). *Pleistocene geology and biology*. Longmans Green: London.

West, R.G. (1969). Pollen analyses from interglacial deposits at Aveley and Grays, Essex. *Proc. Geol. Ass.* **80**, 271–82.

West, R.G. (1972). Relative land-sea level changes in southeastern England during the Pleistocene. *Phil.Trans. R. Soc. Lond.* **A272**, 87–98.

West, R.G. (1980). Pleistocene forest history in East Anglia. *New Phytol.* **85**, 571–622.

West, R.G., Lambert, J.M. & Sparks, B.W. (1964). Interglacial deposits at Ilford, Essex. *Phil.Trans. R. Soc. Lond.* **B247**, 185–212.

West, R.G. & McBurney, C.M.B. (1954). The Quaternary deposits at Hoxne and their archaeology. *Proc. Prehist. Soc.* **20**, 131–54.

Whitaker, W. (1889). The geology of London and parts of the Thames Valley. *Mem. Geol. Surv. G.B.*

Whiteman, C.A. (1987). Till lithology and genesis near the southern margin of the Anglian ice sheet in Essex, England. In: van der Meer, J.J.M. (ed.) *Tills and glaciotectonics*, pp. 55–66. Balkema: Rotterdam.

REFERENCES

Whiteman, C.A. (1992). The palaeogeography and correlation of pre-Anglian Glaciation terraces of the River Thames in Essex and the London Basin. *Proc. Geol. Ass.* **103**, 37–56.

Whiteman, C.A. & Rose, J. (1992). Thames river sediments of the British Early and Middle Pleistocene. *Quatern. Sci. Rev.* **11**, 363–75.

Williams, P.F. & Rust, B.R. (1969). The sedimentology of a braided river. *J. Sedim. Petrol.* **39**, 649–79.

Wilson, T.H. (1897). Note on a section in the Lea Valley at South Tottenham. *Essex Naturalist* **10**, 101–2.

Wintle, A.G. (1981). Thermoluminescence dating of Late Devensian loess in southern England. *Nature* **289**, 479–80.

Wintle, A.G. (1982). Thermoluminescence dating of loess. PACT 6, Proceedings of a specialist seminar on thermoluminescence dating, Oxford, 1980, pp. 486–94. Council of Europe, Strasbourg.

Wiseman, C.R. (1978). *A palaeoenvironmental reconstruction of part of the Lower Thames terrace sequence based on sedimentological studies from Aveley.* M.Sc. thesis. CNAA City of London Polytechnic.

Wood, S.V. (1866). On the structure of the Thames Valley and of its contained deposits. *Geol. Mag.* **3**, 57–63.

Wood, S.V. (1882). The Newer Pliocene Period in England. *Q. J. Geol. Soc. Lond.* **38**, 666–745.

Woodward, H. (1869). The freshwater deposits of the valley of the Lea, near Walthamstow, Essex. *Geol. Mag.* **6**, 385–8.

Woodward, H. (1884). The ancient fauna of Essex. *Trans. Essex Fld. Club* **3**, 1–29.

Woodward, H. & Davies, W. (1874). Notes on the deposits yielding mammalian remains in the vicinity of Ilford, Essex. *Geol. Mag.* **1**, 390–8.

Woodward, H.B. (1890). On the Pleistocene (non-marine) Mollusca of the London district. *Proc. Geol. Ass.* **11**, 335–88.

Woodward, H.B. (1909). The geology of the London district. *Mem. Geol. Surv. G.B.*

Wooldridge, S.W. (1927). The Pliocene history of the London Basin. *Proc. Geol. Ass.* **38**, 49–132.

Wooldridge, S.W. (1938). The glaciation of the London Basin and the evolution of the Lower Thames drainage system. *Q. J. Geol. Soc. Lond.* **94**, 627–67.

Wooldridge, S.W. (1960). The Pleistocene succession in the London Basin. *Proc. Geol. Ass.* **71**, 113–29.

Wooldridge, S.W. & Linton, D.L. (1955). *Structure, surface and drainage in south-east England.* George Philip: London.

Wymer, J.J. (1957). A Clactonian flint industry at Little Thurrock, Grays, Essex. *Proc. Geol. Ass.* **68**, 159–77.

Wymer, J.J. (1968). *Lower Palaeolithic archaeology in Britain.* John Baker: London.

Wymer, J.J. (1974). Clactonian and Acheulian industries in Britain: their chronology and significance. *Proc. Geol. Ass.* **85**, 391–421.

Wymer, J.J. (1977). The archaeology of Man in the British Quaternary. In: Shotton, F.W. (ed.) *British Quaternary studies*, pp. 93–106. University Press: Oxford.

Wymer, J.J. (1985). *Palaeolithic sites of East Anglia.* Geobooks: Norwich.

Zagwijn, W.H. (1984). Sea level in the Netherlands during the Eemian. *Geol. en Mijnb.* **62**, 437–50.

Zeuner, F.E. (1959). *The Pleistocene Period.* Hutchinson: London.

Appendix 1 Pebble counts from high-level gravels in the Epping Forest area

The pebble counts are derived from examination of sieve grades 33, 16 and 8 mm, and are given as percentages.
Grid references for sample sites are given in the text.

Site	Flint			Igneous and metamorphic	Quartz	Quartzite
	Angular	Rounded	Total			
Robin Hood's Pool	66.58	13.52	80.10	0.26	11.73	4.34
Debden Green	76.74	15.76	92.50		4.62	1.03
Buckhurst Hill	83.89	13.67	97.46		1.17	
QE Hunting Lodge	73.45	23.20	96.65		2.32	0.26
Noaks Hill	76.76	15.76	92.50		4.65	1.03
Navestock Side	46.70	49.3	96.0		1.3	1.8
Waltham Forest (N)	36.1	60.6	97.0		1.5	0.3
Waltham Forest (S)	50.6	48.1	98.6		0.5	
Woodford Green	82.05	15.85	97.90		0.91	
Epping Forest Great Monk Wood	66.22	13.55	79.77	0.19	11.83	2.67

					Chert					
Sand-stone	Calcar-eous sand-stone	Brown sand-stone	Green-sand	Chalk	Green-sand chert	Grey chert	Rhaxella chert	Other chert	Schorl	Total
0.26		0.26	1.59		2.04	0.77				392
							1.81			387
		0.40			0.59	0.20			0.20	512
		0.52								388
					0.57					387
		0.4			0.4					227
	0.3	0.6				1.3		1.3		396
	0.3					0.6		0.6		631
		0.21		0.98						488
0.95		0.95			2.29	0.57	0.19		0.57	524

Appendix 2 Pebble counts from the Lower Thames region

The pebble counts are derived from examination of sieve grades 33, 16 and 8 mm, and are given as percentages.
Grid references for sample sites are given in the text.

Site	Flint angular	Flint rounded	Flint total	Igneous and metamorphic	Quartz	Quartzite
Dartford Heath Gravel						
Buckingham Hill, Mucking	49.8	39.6	89.4		3.8	4.3
Dartford (Brooklands Rd)	52.5	39.2	91.7		3.5	1.7
Dartford (Coldblow Wood)	62.7	25.1	87.9		4.0	3.1
Dartford Heath	56.5	30.7	87.2		5.0	3.4
Dartford Heath Cottages	68.6	20.8	89.4		2.9	1.4
Dartford Tunnel App S (3.4–3.6 m)	31.5	61.6	93.0		0.8	0.4
Dartford Tunnel App S (CT)	42.7	49.0	91.7	0.4	1.4	1.2
Dartford Tunnel S App (2 m)	39.0	51.0	90.0		2.0	1.3
Southfields, 1.5 m	44.8	46.3	91.1		3.2	2.3
Southfields, 2.5 m	55.8	34.8	90.7		2.8	2.5
Swanscombe Member						
Swanscombe (7 m)	59.6	31.9	91.5		2.6	1.5
Swanscombe (Galley Hill)	73.7	18.5	92.3		1.6	0.9
Swanscombe L.G.	41.2	38.9	80.1		9.2	4.0
Swanscombe L.M.G.	71.0	19.1	90.1		3.3	1.6
Orsett Heath Gravel						
Barnehurst (Manor House)	66.8	23.2	90.0		3.5	1.6
Chadwell St Mary	49.9	41.1	90.9		2.6	1.6
Hornchurch Cutting	50.9	41.8	92.6		0.5	3.7

Sand-stone	Calcar-eous sand-stone	Feldspar	Green-sand	Chalk	Chert Green-sand chert	Grey chert	Rhaxella chert	Other chert	Total
					2.0	0.2	0.2	2.4	444
0.4		0.2			2.3	0.2		2.5	482
	0.2				4.6		0.2	4.8	453
0.3	1.1				2.6	0.2			616
0.5		0.5			3.6	1.8		5.4	442
					5.5	0.2		5.7	473
0.4	0.6	0.2	0.4		3.7		0.2	3.9	518
	0.3				6.4			6.4	752
0.4	0.6				1.7			1.7	473
	0.5	0.5	0.5	2.5			2.5		396
		0.2	3.4	0.8		4.2	532		
0.9					4.2	0.2		4.4	426
	0.2			2.4	2.5	1.1		3.6	447
0.2	0.2			4.3	0.5	0.2	5.0		634
0.3					4.2	0.4		4.6	738
0.4	0.2	0.2			3.0	0.3	0.3	3.6	607
					2.3	0.3	0.6		362

APPENDIX 2

Site	Flint			Igneous and metamorphic	Quartz	Quartzite
	angular	rounded	total			
Horns Cross (3 m)	57.4	31.1	88.5		3.7	2.7
Moor Hall, Aveley (2 m)	52.0	35.7	87.7		7.9	0.9
Moor Hall, Aveley (3.5 m)	49.9	34.3	84.1		6.6	4.3
St Paulinus, Crayford	54.5	34.2	88.7		4.1	1.8
South Orsett (3 m from surface)	52.9	38.2	91.1		3.5	1.5
Stifford (Upper pit 2, 1.5 m)	39.6	48.6	88.1		4.6	2.1
Stifford (Upper pit)	47.2	42.1	89.2		5.5	0.9
Stifford (Warren Lane)	48.7	28.3	87.1	0.6	6.3	1.2
Stone Cross (1 m)	39.9	46.3	86.3		7.5	1.0
Stone Cross (2 m)	63.7	22.5	86.2		5.2	2.5
Swanscombe (Crayland's Ln)	55.4	35.2	90.5		3.7	2.5
West Tilbury	42.9	46.0	88.9	0.2	3.3	2.0
Corbets Tey Gravel						
Arena Essex (1.5 m)	74.8		90.6		4.3	1.3
Arena Essex (3 m)	64.2	23.8	88.0	0.2	6.3	2.0
Arena Essex Site 3 (1 m)	89.0	20.0	89.0		4.7	3.8
Aveley (Love Lane)	73.5	15.3	88.8		5.5	1.3
Aveley (Marleys E)	61.1	28.6	89.7		5.5	1.9
Aveley (Marleys W, 1 m)	61.6	27.7	89.3		4.4	3.2
Aveley (Marleys, 3.5 m)	54.1	30.7	84.8	0.1	7.4	2.7
Barvils Farm	62.1	32.3	94.3		5.7	
Bush Farm (1.5 m)	59.4	29.0	88.4	0.1	5.9	3.1
Bush Farm (1 m)	62.8	25.4	88.3		5.5	1.9
Corbets Tey	53.6	31.5	85.0		7.5	3.0
Fairlop (2 m)	77.5	11.6	89.0		3.7	1.7
Fairlop (4 m)	67.8	23.3	91.1		4.8	2.0
Gerpins Lane, Rainham	62.0	25.4	87.5		4.6	2.3
Grays Thurrock (Botany)	62.1	23.9	86.0	0.4	6.2	3.7
Mucking (1 m above base)	67.5	25.4	92.9		2.6	0.8
Mucking (St Cleres)	39.4	47.7	89.1	0.2	3.4	2.6
N. Rainham (0.5 m)	69.8	20.8	90.6		2.6	1.3
S Romford (Crown Fm)	67.7	22.4	90.1		6.1	1.7
Snaresbrook Park, Leyton	50.8	40.4	91.2		4.8	1.4
South Ockenden	74.2	14.7	89.0	0.2	4.8	1.8
Stifford (Sect 1, 3 m)	87.5	6.6	94.0		1.2	1.0
Stifford (Sect 4, 2 m)	66.5	20.8	87.3		4.8	2.4
Stifford (Swallow Hollow)	75.9	15.7	91.7		4.0	1.4
Tilbury (Low Street)	63.7	28.3	92.0		4.3	0.6
Wanstead (Station)	75.4	15.2	90.6		4.5	2.3

PEBBLE COUNTS FROM THE LOWER THAMES REGION

Sand-stone	Calcar-eous sand-stone	Feldspar	Green-sand	Chalk	Chert				Total
					Green-sand chert	Grey chert	Rhaxella chert	Other chert	
	1.5				2.8	0.8		3.6	1064
	0.2	0.5			2.1	0.8		2.9	633
0.1	0.7				3.2	0.4		3.6	692
			0.2		4.1	0.8	0.5	5.4	664
	0.2				3.3	0.2		3.5	518
1.9				2.1	0.4	0.3	2.8		675
	0.2				3.6	0.2	0.2	4.0	530
0.4		2.2			1.2	1.2		2.4	511
					5.0	0.2	5.2		604
					5.1	0.7		5.8	689
0.2	0.4				2.5			2.5	516
0.4					4.5	0.4		4.9	552
	0.2					0.2		3.6	445
0.2	0.3	0.2	0.2		2.3	0.3		3.6	662
0.2					2.0	0.2	0.2	2.4	451
					4.0	0.3	0.3	4.6	400
	0.2	0.2			1.7	0.6		2.3	476
	0.2				2.3	0.3	0.2	2.8	656
0.2	0.2				4.0	0.6		4.6	525
0.2									459
0.1	0.1				1.8	0.3		2.1	717
0.3	0.1				3.4	0.3	0.3	4.6	732
0.4					2.6	0.4		3.0	534
0.2					5.2		0.2	5.4	520
					1.7	0.2		1.9	459
	0.2	0.2	0.2		5.0			5.6	519
			0.4		3.3			3.3	520
			0.2	3.8					495
0.2		10			4.6	0.6	0.2	5.4	497
			1.3		3.6		0.2	3.8	533
					2.0			2.6	344
			0.2		2.1			2.1	421
0.3	0.3				3.4	0.5		3.9	625
					3.0	0.6		3.6	503
0.2					5.0			5.0	418
			0.2		2.4	0.4		2.8	503
					2.7	0.4	3.7		486
			0.2		2.3	0.2		2.5	487

APPENDIX 2

Site	Flint			Igneous and metamorphic	Quartz	Quartzite
	angular	rounded	total			
Mucking Gravel						
Becontree Heath	75.3	14.9	90.1		3.9	1.4
East Court, Chalk (2 m)	6.6	91.5	97.1			
Ilford Brook (1.5 m)	66.0	27.9	93.9		3.1	0.5
Mucking (1 m below top)	66.7	23.6	90.2		3.1	0.7
Mucking (Collingwood Fm)	58.5	29.6	88.6		3.3	2.4
Rush Green Dagenham	72.2	18.4	90.6		2.9	1.8
South Hornchurch	39.9	45.3	85.3		10.2	2.9
Tilbury (Gravel Pit Farm)	56.6	34.8	91.4		2.7	2.5
Wanstead (Bush Road)	62.3	22.7	85.6		7.0	4.0
Wanstead Flats Pond (1.5 m)	66.8	24.1	90.2	0.2	4.6	2.6
West Thurrock Gravel						
W Thurrock (Lion Cutting)	31.5	59.5	91.0		6.8	1.4
East Tilbury Marshes Gravel						
E. Tilbury (1 m above base)	57.7	33.4	91.1		3.4	2.2
E. Tilbury (1 m from top)	60.8	32.3	93.2		2.1	1.0
Stone Marsh (3 m)	52.5	34.8	89.3	0.2	6.9	1.7
Others						
Barnes Clay (1.5 m)	24.2	74.9	99.1	1.1	0.3	
Greenhithe	44.4	43.2	87.6			0.3
Nightingale Estate 1	75.4	19.5	94.9		0.4	1.8
Nightingale Estate 2	78.3	14.8	93.4		0.7	2.9
Purfleet (basal gravel)	55.5	30.8	86.3	0.1	6.4	3.0
Stamford Hill	78.5	10.6	89.1	0.2	4.7	2.5

PEBBLE COUNTS FROM THE LOWER THAMES REGION

Sand-stone	Calcar-eous sand-stone	Feldspar	Green-sand	Chalk	Chert				Total
					Green-sand chert	Grey chert	Rhaxella chert	Other chert	
0.2	0.2		0.2		3.6	0.2		3.8	558
	1.7				1.2			1.2	484
0.7					1.7		0.2	1.9	423
			0.2		5.3				543
0.6		0.6	0.9		3.9	0.3		4.2	337
					3.5	0.2		3.7	668
					1.1	0.3	0.3	1.7	373
					2.7	0.6		3.3	489
			0.2		3.6	0.2		3.8	472
					3.1	0.2			569
					0.9			0.9	222
		1.1			2.2			2.2	553
			0.2		3.4		0.2	3.6	526
					3.5	0.4		3.9	480
			0.3		0.3			0.3	359
	0.4	0.3			11.7			11.7	707
					2.5		0.2	0.2	786
					1.9		0.2	0.9	961
0.5	0.4	0.2			3.1			3.1	708
	0.6				1.8	0.4		0.7	565

221

Index

A102(M)
　Blackwall Tunnel section 96
A13 borehole data 67, 69
　Corbets Tey Member 38
　Orsett–Rainham 22–3
　Orsett–Tilbury link, section 107
Acheulian artefacts
　Early Middle, hand-axes 27–8
　Late Middle 56
　stratigraphic distribution 164–5
　see also Artefacts
Alluvium
　defined 114
Amino acids
　racemisation data 174–7, 182, 185
Anglian Stage
　Lowestoft Formation 121
　palaeogeographical evolution 178–81
　sequence of events 175
Appendices
　Lower Thames Valley 214–21
Arctic Bed, Lea Valley 109–10, 192–3
　vertebrate faunas 133, 136
Ardleigh Green
　borehole data 32
Artefacts 162–73
　classification (Wymer) 163
　incorporation into sediments 163–6
　relationship to stratigraphic units 164–5
　sites
　　Belhus Park, Aveley 172
　　Bluelands, Purfleet 172
　　Clactonian 166–9
　　Corbets Tey Gravel 169–70
　　Crayford–Erith area 171
　　Dartford Heath Gravel 167
　　Grays Thurrock 171
　　Greenlands, Purfleet 172
　　Ilford 172
　　Ingress Vale 168
　　Lea Valley Gravel 173
　　Mucking Gravel 170
　　Northfleet 170–1
　　Orsett Heath Gravel 169
　　Stoke Newington 172
　　Swanscombe Member 167–8

　　Thurrock, West 171
　　Wolstonian stage 183–4
Aurochs see Bos/Bison
Aveley
　Aveley–West Thurrock–Crayford silts and sands 59–88
　Belhus Park 74–5
　Camden Town 86
　Crayford–Erith 69–72
　Grays Thurrock 61–4
　Hackney Downs, Stoke Newington and Highbury 80–5
　Ilford 77–80
　Northfleet 59–61
　Ockendon, North, and Upminster 75–6
　Peckham 86–8
　Purfleet 73–4
　Rainham sewer 89
　Sandy Lane Quarry 68–9
　sequence 60
　West Thurrock 64–8
Aveley's Pits, M25 36
Belhus Park
　artefacts 172
　M25 cuttings 74–5
　palaeobotany 146–9
　sediments and sequences 74–5
Ipswichian 187
Love Lane 38
Marley's Pit, Silt Complex 98
Moor Hall 31
Mucking gravel 54
Sandy Lane Quarry 68–9, 98
　palaeobotany 145–6
vertebrate faunas 132–3, 138, 139

Baker's Hole Industry
　British Museum site 60–1
　Northfleet 59–61
Baldwin's Farm Quarry, North Ockendon 34, 75–6
　palaeobotany 149, 150
Barking area
　Barking Creek 112
　Castle Green 91
　Fairlop–Rush Green spread 41
　Mucking Gravel 54

Upton Park 55
Barking Marshes 104
Barling Gravel 199
Barnehurst
　Manor House 31
Barnfield Pit 28, 181
　bedrock collapse 182
　Swanscombe Member deposits 25–6
　vertebrate faunas 137
Barvill's Farm
　Corbets Tey Gravel 39
Beam, river 54
Becontree Heath 54
Becton, river 104
Belhus Park see Aveley
Bethnal Green
　Mucking Gravel 56
Black Park Gravel, Middle Thames 19, 180
　artefacts 167
　correlation with Dartford Heath Gravel Member 19
Blackfriars
　deposits 56
Blackfriars Bridge 117
　Blackfriars–Camberwell, section 95
　Blackfriars–Shadwell, borehole data 49
　Fleet confluence 102
Blackheath Beds
　flints 16
　pebble beds 20
　principal components analysis (PCA) 121
Blackwall Tunnel
　borehole data 117–18
　Greenwich–Isle of Dogs 96
　Shepperton Gravel 102
Blackwater Estuary 199
Blade Park Gravel
　correlations 196
Bluelands Pits see Purfleet
Borehole data
　location map 2–3
　sources 12
Bos (aurochs) 25
Bos/Bison 132–3, 135, 136, 138

222

INDEX

Botany Pit, Purfleet 38, 74
Bow
    Old Ford, borehole data 32
Bowman's Lodge Pit 19
Boyn Hill Gravel Member
    correlations 30, 32, 196, 197
    Havering 32
    Herne Hill 32
    Hornchurch 32
    Islington 32, 33
    Middle Thames, artefacts 167
    see also Orsett Heath Gravel
Boyn Hill Terrace 19, 20, 29–30
'Brickearth'
    artefacts 171
    Crayford–Erith area 52, 69–72
    Globe Pit 62
    Langley Silt Complex 94–7
    Middle Loam 60, 62
    vertebrate faunas 138–9
    see also Silt Complex
British Museum site
    Baker's Hole Industry 60–1
Bromley
    Lea valley 102
Brooklands Road pits 19–20
Broomfield House
    Corbets Tey Gravel 49
Buckhurst Hill 14
Buckles Farm
    South Ockendon 34
Buried Channel, Lower Thames
    Valley 101
    see also Shepperton Gravel
Camberwell
    Camberwell–Blackfriars section 95
    Denmark Hill 88
    Earl's Sluice (lost river) 102
Camden Town
    Aveley–West Thurrock–Crayford
        silts and sands 86
Canis sp. 27
Canning Town
    Lea confluence 116
Celcon Pit, Grays Thurrock 52, 61
Central Line
    Holborn–Stepney, section 57
Chadwell Place
    Mucking Gravel 52
Chadwell St Mary
    Orsett Heath Member 30
Charlton
    New, Shepperton Gravel 102
Chigwell
    sediments 18, 179
Clactonian artefacts 25, 26, 62, 74, 166–9
    Flakes, Globe Pit 39
    stratigraphic distribution 164–5
    see also Artefacts
Clapton Common
    borehole data 49
Clapton Railway Station 82
Claygate Beds 14
Clayhall
    Fairlop–Rush Green spread 41

Clerkenwell–Shoreditch deposits 43, 48
Coelodonta 132–3, 135
    see also Vertebrate fauna
Collingwood Farm
    Dartford Heath Gravel 20
Colne, river 180
'Coombe rock' 59–61, 73, 126–7
Copenhagen Fields 86
Corbets Tey Gravel Member 30, 32, 34–49
    artefacts 169–70
    deposition 183–4
    extension, correlations 199
    gradient 43
    Langley Silt Complex 98
    long profile 35, 50
    pebble lithography 43
    principal components analysis (PCA) 120–2
    sites
        A13 borehole data 38
        Barvill's Farm 39
        Broomfield House 49
        Bush Farm 98
        Corringham 39
        Hackney 58
        Lea Valley 56
        Purfleet 73
        Purfleet–Ockendon area 38
        Snaresbrook Park 41
        Stanford-le-Hope 39
        stratotype 34
        Taplow-equivalent Mucking Gravel 51
    vertebrate faunas 132–3, 135
Corbicula bed
    Crayford–Erith area 70
    freshwater origin 186, 187
    Ilford sequence 78
    Uphall Pit sequence 77
Corringham
    Corbets Tey Gravel Member 39
Cranham Church
    South Ockendon 34
Cranham Hall
    Upminster 76
Cray, river 112–14
    confluence 20
    Shepperton Gravel 104
    Cray Valley 20
    Mascal Hill 31
Crayford silts/sands see
    Aveley–West Thurrock–Crayford silts and sands
Crayford–Erith area
    artefacts 171
    Aveley–West Thurrock–Crayford silts and sands 69–72
    'brickearth' 52, 69–72
    Corbicula bed 70
    East Tilbury Marshes Gravel, borehole data 90–1
    Erith to Slade Green, section 72
    Mucking Gravel 52

sites
    Crayford Road 72
    Furner's Old/New Pit 69
    Myrtle Close 70–1
    Norris' Pit 69
    Northend, sequences 71–2
    Perry Street 71
    Rutter's Pit 69
    St Paulinus' Church 31
    Stoneham's Pit 69
    Talbot's Pit 69
    Tertiary-type beds 71
    Thanet Sand 69
    vertebrate faunas 133, 135, 138
    Woolwich beds 70–1
    see also Aveley–West Thurrock–Crayford silts and sands
Crossness
    Tilbury beds 116

Dagenham
    District Line 54
    Reede Road 54
    Tilbury biogenic beds 116
Dama dama (deer) 25, 132–3, 136, 138
Darent, river 112–14, 178–9
    confluence 20
    Shepperton Gravel 104
Darent Valley
    Dartford Heath Gravel 19
    deposits 15, 16
    Hawley Gravel 113–14
    M25 section 113
    Sutton Place 113
Darent–Cray, river 178–9
Darenth Wood Gravel 14, 15
    correlations 198
    palaeogeographical evolution 178
Dartford
    Tilbury biogenic beds 115
Dartford Heath Gravel Member 18–24
    artefacts 167
    correlation with Black Park Gravel 19
    erratics 121
    gradient 23
    incision event 180–1
    principal components analysis (PCA) 120–2
    profile 24
    Rhaxella chert 23
    sites
        Bowman's Lodge 99
        Bowman's Lodge Pit 19
        Coldblow Wood 20
        Darent Valley 19
        Gravel Pit Cottages 19–20
        Orsett Pit 30
        Wansant Pit 18, 19
    vertebrate faunas 132–3, 135
Dartford Marshes
    Long Reach Hospital 104
    Shepperton Gravel 104
Dartford Tunnel see M25
Debden Green Gravel 13, 14, 178

INDEX

Denmark Hill
    Camberwell 88
Dennis's Cottages
    Ockendon, North 34, 75–6
Deptford
    Ravensbourne confluence 102
Devensian Stage
    palaeogeographical evolution 191–3
    sequence of events 175
Diamictons 126–7
Dicerorhinos hemitoechus 77, 132–3, 137
District Line
    Dagenham 54
Dollis Hill Gravel 197
Donjek depositional model 126

Earl's Sluice (lost river)
    Camberwell 102
East Cross Route
    Hackney–Redbridge–M11 42–3, 94, 111, 118
East Ham
    Gallions Sewage Works 104
    M15 route, section 94
    Roding river 104, 112
East Tilbury Marshes Gravel
    and Aveley Silts/Sands 69
    Devensian Stage 192
    Kempton Park 89–94
    and Leyton Gravel 94
    long profile 92
    Lower Thames Valley 102
    vertebrate faunas 132–3, 135, 136
    see also Tilbury
Ebbsfleet Valley
    Ipswichian 188
    Wolstonian 184–5
Eemian Stage
    sea level 189
Effra Gravel 14
Effra Terrace 14
Elephas primigenius fauna
    West Thurrock 171
Epping Forest Formation 122
    Epping Forest Members 195
    interfluve 13–15
    palaeogeographical evolution 178
    pebble counts 214–15
Equus ferus (horse) 25, 27, 132–3, 135, 136
Erith
    Mucking Gravel 52
    to Slade Green, borehole data 72, 90–1
    see also Crayford–Erith area
Erith Marshes
    section 103, 104
Erosion surfaces 125, 126
Essex
    Central, Kesgrave Formation 195–7, 198
    Mid-Essex depression 195
    South-eastern, correlations 198–200

Esso Oil Terminal
    Purfleet 74

Facies assemblage studies
    gravel members 123–30
Fairlop Gravel 30, 39
    Ilford sediments 98
Fairlop–Gants Hill
    Rom–Roding interfluve 32
Fairlop–Rush Green spread 41
Faunal assemblages 131–9
Finsbury–Roding Valley
    M11 section 44
    Mucking Gravel 56
Flandrian Stage
    palaeogeographical evolution 193–4
Fleet, river
    confluence, Blackfriars Bridge 102
    'lost river' 107–9
        Camden Town 86
    valley 109, 117
        Corbets Tey Gravel 43
        Taplow Gravel 56
Floodplain gravel
    carbon dating 90
    first use of term 89–90
    upper and lower 90
Floodplain Terrace 43, 51
Formation
    defined 11
Furner's Old/New Pit, Crayford–Erith area 69

Galley Hill Pit 27
Gallion's Sewage Works 116
Gants Hill
    Fairlop–Rush Green spread 41
Geological map 6–7
Gidea Park borehole data 32
Gironde Estuary
    Flandrian 190
Glacial gravel and till 16–18
Globe Pit, Grays Thurrock
    'brickearth' 62
    Corbets Tey Gravel 39
    fossils 63
    Mucking Gravel 52
Gravel units
    sedimentary structures 123–6
Gravesend
    Allhallows 107
    Rosherville Works 106
    Shornmead Fort 107
    Terrace Pier 106–7
Grays
    A borehole data, pollen counts 141
    Chalk Quarry 31
    East Tilbury Marshes Gravel 90
    Globe Pit 39, 52, 62, 63
    Orsett Heath spread 31–2
'Grays brickearth' see 'Brickearth'
Grays Thurrock 52, 61–4, 89
    artefacts 171
    borehole data 63–4
    Celcon Pit 52, 61

Globe Pit 52
    Ipswichian 186–7
    mammal remains 61
    Mollusca 63
    quarry, sections 62
    vertebrate faunas 132–3, 139
    see also Thurrock
Great Monk Wood
    Rhaxella chert 18
    Robin Hood's Pool Gravel 179
Greenhithe chalk quarry 16
    see also Purfleet
Greenwich
    Greenwich–Isle of Dogs 92
        Blackwell Tunnel section 96
    National Maritime Museum 92
Group
    defined 11
Gun Hill
    Orsett Heath Gravel 31

Hackney
    M11 borehold data 41
    Mucking Gravel 56
    South
        to Islington, section 45
        to Shadwell, section 46
Hackney Brook
    confluence with Lea 58
    valley 58
Hackney Downs
    Arsenal area 56–8
    borehole data, section 47
    Hackney Downs Gravel 83, 85
    Langley Silt Complex 98
    Nightingale Estate 83–4, 188–90
    palaeobotany 153–5
    organic sediments 43, 85, 188
    Stoke Newington and Highbury 81
        Aveley–West Thurrock–Crayford silts and sands 80–5
    palaeolithic floors 81–2
Hackney–Redbridge
    East Cross Route, sections 43, 94, 111, 118
Hangman's Wood
    M25 38
Havering
    Boyn Hill Gravel 32
Hawley Gravel
    Darent Valley 113–14
Hays Wharf
    London Bridge 102
Herne Hill
    Boyn Hill Gravel 32
'Herring-bone' cross-bedding 130
High Beach Gravel 15, 178
High-level deposits 12–18
    glacial gravel and till 16–18
    gravels 13–16
    Greenhithe chalk quarry 16
Highbury
    New Park
        borehole data 84
        brickpits 81

palaeobotany 155–8
  to Islington, section 58
Highbury Silts and Sands Member
  83–5
Hippopotamus 132–3, 135–6, 138–9
Holborn
  Corbets Tey Gravel 43
  Holborn–Whitechapel, Silt
    Complex 97
Holborn Viaduct
  Fleet valley 109
Holborn–Stepney
  Central Line, section 57
Homerton
  Mabley Green 111
Hornchurch
  Boyn Hill Gravel 32
  South, borehole data, section 54
Hornchurch Till 16–18, 178–9
  Orsett Heath Member 183
Horns Cross 31
Howbury Pit, Slade Green 89
Hoxnian Stage 30
  sequence of events 175
Hoxnian–early Wolstonian Stages
  palaeogeographical evolution
    181–3
Hunts Hill
  Mucking Gravel 52
Hydrobia
  Ipswichian 187

Ice wedges 125, 126
Ilford
  artefacts 172
  Corbicula bed 78
  Late Middle Acheulian artefacts
    56
  Mucking Gravel 54, 55
  palaeobotany 149–53
  Roding Valley Gravel 112
  sites
    Connaught Road 78
    Gordon Road 78
    Green Lane 78
    High Road 77–80
    Langley Silt Complex particle-
      size 99
    Riches (Richmond) Road 80, 99,
      152
    Seven Kings Station 79, 80
    Uphall Pit 77, 133, 137–9
    Winston Way 54, 55, 79–80
  three units 80
  town centre, sections 55
  vertebrate faunas 133, 137–9
Ingrebourne, river 54
Ingress Vale 28, 181
  artefacts 168
  vertebrate faunas 132–3, 137
International Stratigraphic Guide
  nomenclature 11
Ipswichian Stage
  palaeogeographical evolution
    185–91
  sequence of events 175

Iron oxide deposits 15
Isle of Dogs–Greenwich 92
  section 96
  Tilbury beds 116
Islington
  Boyn Hill Gravel 32, 33
  to Highbury, section 58
Islington–South Hackney
  section 45
Isoleucine
  D-alloisoleucine/L-isoleucine
    ratios, Mollusca 177

Kemps
  Ockendon, North, borehole data
    75–6
Kempton Park Gravel Member 90
  correlation with East Tilbury
    Marshes Gravel 89–94, 196
  and Silt Complex 98
  South Thamesside 92
Kesgrave Formation
  Central Essex 195–7, 198
Kingston Gravel 18

'Lag' gravel 27
Langley Silt Complex
  Devensian 192
  origin 97
  particle-size distribution 100
  see also Silt Complex
Lea Leytonstone Gravel 51, 83, 85
Lea, river 54, 109–12
  confluence
    Canning Town 116
    with Hackney Brook 58
    Royal Albert Docks 102
    Royal Victoria Docks 102
    Shepperton Gravel 116
    with Thames 110
    Thames, West Ham 117–18
  floodplain 117–18
    Leyton Gravel 94
  Ipswichian 189
  Limehouse–South Bromley 92
  palaeographical evolution 180
  Pickett's Lock 109–10
Lea Valley
  Arctic Bed 109–10, 192–3
    vertebrate faunas 133, 136
  Bromley 102
  Corbets Tey Gravel equivalent 49
  correlations 51
  floodplain alluvium 117–18
  Hackney Brook tributary 85
  Leyton Gravel 51
  Leytonstone Gravel 51, 56
  Mucking Gravel 56
  Stamford Hill Gravel 51
  terraces, carbon dating 110
Lea Valley Gravel 51, 111, 117–18
  artefacts 173
  Low Terrace 112
  pebble lithography 112
  Tottenham Hale 118
  Victoria Line 118

Lea–Roding interfluve 39, 41
  Mucking Gravel 55
  Wanstead 32
Lemmus sp. 27, 132–3, 136
Levalloisian artefacts 64, 69, 74
  classification 163
  stratigraphic distribution 164–5
  see also Artefacts
Leyton Gravel
  floodplain, Lea 94
  section 51, 192
Leytonstone Gravel 51, 83, 85, 184
  borehole data 56–9
  Corbets Tey Gravel 49
Leytonstone Hospital 41
Limehouse Reach
  Shepperton Gravel 102
  Tilbury 117
Lion Tramway Cutting
  West Thurrock 54, 64
  WTA and WTF pollen counts 143
Lithostratigraphy
  nomenclature 9–10
Liverpool Street Station 56
Location map 2–3
Loess
  'Brickearth' 61
  origins of salts (loess pupchen)
    96–7
London Bridge
  Hays Wharf 102
  to Peckham, sections 87–8
  Walbrook confluence 102
London Docks 116, 117
  Shepperton Gravel 102
Loughton Gravel 18
Love Lane
  Aveley 38
Low Street
  Mucking Gravel 52
Lower Floodplain Gravel 100–1
  Swanscombe Member deposits
    25–6
Lower Loam
  floodplain 181
  Swanscombe Member deposits
    26
Lower Middle Gravel 27–8, 181–2
Lower Thames Valley
  appendices 214–21
  artefacts 162–73
  Buried Channel 101
  correlations 195–200
    East Tilbury Marshes Gravel
      89–94, 102
  flood protection 114–15
  floodplain
    gravels 107–14
    width 114–15
  geological background 4
  geological maps 6–7
  lithostratigraphy 118
  location map 2–3
  London Bridge, Walbrook
    confluence 102
  palaeobotany 140–61

palaeogeographical evolution 174–94
  reconstructions 176
  sequence of events 175
pebble lithography 119–22
  pebble counts 214–15
previous investigations 4–8
references 201–13
sediments and deposits 123–30
terrace nomenclature 9–10
terrace stratigraphy 8–9
Tilbury biogenic beds 115
topography 4
vertebrate faunal assemblages 131–9
Lowestoft Till Formation 16–18
Lynch Hill Gravel
  correlation 43, 196
Lynch Hill Terrace 34

M11 motorway
  borehole data 41–3
    Mucking Gravel 56
  East Cross–Hackney 111
  East Ham–Barking area 112
  Hackney link (East Cross Route), 43, 94
  junction, North Circular Road, borehole data 44
  section, Finsbury–Roding Valley 44
M12 motorway
  borehole data, section 39, 40
M15 extension
  Plumstead–Erith, section 105–6
  route section, East Ham, Newham 94
M25 motorway
  and A13 (Sandy Lane), West Thurrock 67
  Aveley 74–5
  Aveley's Pits 36
  borehole data 21, 36, 37
  Dartford Tunnel 64, 106
    borehole data 21
    East Tilbury Marshes Gravel 90
    Tilbury biogenic beds 115
  Hangman's Wood 38
  Ockendon North 75–6
  Purfleet–Cranham 36, 37
  section, Darent Valley 113
  Thanet Sands 38
Mammuthus primigenius 132–3, 134, 137, 138–9
  Aveley 68, 71, 77
Manor Farm
  South Ockendon 76
Maps of Lower Thames Valley
  geological map **6–7**
  location map **2–3**
Mar Dyke valley
  Corbets Tey Gravel 38
  Ipswichian 190–1
  Ockendon, South 76
Marley's Pit, Aveley 98

Mascal Hill
  Cray Valley 31
Mass flow deposits
  sedimentary structures 126–7
Medway, river 16
Middle Loam 60
Middle Terrace
  Lower Thames valley 49
Middle Thames Valley
  correlations 197–8
  Spring Gardens Gravel 185, 196
Middle and Upper Gravel
  Swanscombe Member deposits 26–9
Mole–Wey valley 179, 180, 197
Mollusca
  Baker's Hole 61
  D-alloisoleucine/L-isoleucine ratios 177
  Grays Thurrock 63
  Hoxnian Stage 181
  Ipswichian 187
  Northfleet 189
  Rhenish element 182
Moor Hall
  Aveley 31
Mucking Gravel 31, 49–58
  artefacts 170, 184
  deposits 53
  long profile 53
  pebble lithography 56
  principal components analysis (PCA) 120–2
  sites
    Chadwell Place 52
    eastern Essex 199
    Hunts Hill 52
    Lea–Roding interfluve 55
    Low Street 52
  and Taplow Gravel 43
  type section 51
  vertebrate faunas 132–3, 135, 184
Musk-ox *see Ovibos*

Navestoke Side 16
Nekinger
  lost river 92
Newham
  Late Middle Acheulian artefacts 56
  Newham–East Ham, section 94
  Newham–Hornchurch, East Tilbury Marshes Gravel 91
  Northern Outflow Sewer, section 92, 93
Nightingale Estate
  Hackney Downs 83–4, 188–90
  palaeobotany 153–5
Nomenclature
  binomial system 9–10
  International Stratigraphic Guide 11
Norris' Pit, Crayford–Erith area 69
North Circular Road
  M11 junction, borehole data 44
North London Line 43

Northern Outfall Sewer
  Newham, section 92, 93
  West Ham 111
Northfleet
  artefacts 170–1
  Baker's Hole Industry 59–61
  Ipswichian 188
  Mollusca 189
  and Swanscombe sequence 24, 181
  vertebrate faunas 132–3
Norwood (Crystal Palace) Terrace 14
Norwood Gravel
  correlations 178

Ockendon
  Purfleet–Ockendon area 38
Ockendon Channel 76
Ockendon, North
  Baldwin's Farm Quarry 75–6
  Baldwin's Pit 34
  Dennis's Cottages 75–6
  Ipswichian 186–7
  Kemps 75–6
  Langley Silt Complex 98
  M25 borehole data 75–6
  palaeobotany 149
  Pea Lane 75–6
  Stubbers Youth Camp 75
  and Upminster, Aveley–West Thurrock–Crayford silts and sands 75–6
Ockendon, South
  Buckles Farm 34
  Cranham Church 34
  Gerpins Lane 34
  Manor Farm 76
  Mar Dyke valley 76
Old Ford
  Bow 32
  section 56
Ongar
  sediments 179
Orsett Heath Gravel Member 39, 179
  artefacts 169
  correlations 35
  long profile 35
  origin 32
  pebble lithography 34
  principal components analysis (PCA) 120–2
  sites 29–34
    Gun Hill 31
    Hornchurch Till 183
    North Stifford 38
    Orsett Cock p.h. 20, 198
    Orsett–Rainham, A13 borehole data 22–3
    Stone Cross 20
  vertebrate faunas 132–3, 135
Ovibos (musk-ox) 28, 132–3, 138
Ox *see Bos/Bison*
Oxygen Isotope Stage 182

Paladihia
  Ipswichian 187
Palaeobotany 140–61

INDEX

aquatic plants 151, 153, 155, 160
Aveley
  Belhus Park 146–9
  Sandy Lane quarry 145–6
  Grays Thurrock 140–2
  Highbury 155–8
  Ilford 149–53
  Nightingale Estate, Hackney 153–5
  North Ockendon 149
  Peckham 158–61
  Purfleet 146–7
  West Thurrock 143–5
Palaeogeographical evolution 176
  Anglian Stage 178–81
  Darenth Wood Gravel 178
  Devensian Stage 191–3
  Epping Forest Formation 178
  Hoxnian–early Wolstonian Stages 181–3
  Ipswichian Stage 185–91
  Lower Thames Valley 174–94
  Warley Gravel 178
Palaeolithic artefacts *see* Artefacts
Palaeoloxodon antiquus 132–3, 134, 137, 139
  Aveley 68
Park Corner Farm
  Upminister 76
Pebble Gravel 13, 16
Pebble lithography
  Epping Forest 214–15
  Lower Thames Valley 216–21
  methods 12
  principal components analysis (PCA) 119–22
Peck, river 92
Peckham
  Aveley–West Thurrock–Crayford silts and sands 86–8
  borehole data 86
  Harder's Road, pollen counts 159
  late-Glacial infill 186
  palaeobotany 158–61
  Queen's Road 88
  regional setting 86–8
  Spring Gardens Gravel 185
  to London Bridge, sections 87–8
Pickett's Lock
  Ponder's End 109–10
Pisidium sp. 181
Plants *see* Palaeobotany
Plumstead
  Southern Outfall Sewer, section 104
Plumstead Marshes
  section 103, 104, 105–6
Pollen counts *see* Palaeobotany and specific sites
Ponder's End
  Pickett's Lock 109–10
Ponder's End Stage 193
Poplar Fire Station
  Shepperton Gravel 102
Principal components analysis (PCA)
  pebble lithography 119–22

Purflet
  anticline 32
  breach 178
  Aveley–West Thurrock–Crayford silts and sands 73–4
  Bluelands Pits 38, 73–4
    artefacts 172
    palaeobotany 147
  Botany Pit 38, 74
  chalk 64, 69
  Chalk bedrock anticline 104
  Corbets Tey Gravel 73
  Esso Oil Terminal 74
  Greenlands Pits 38, 73–4
    artefacts 172
    palaeobotany 147
  palaeobotany 146–7
  Purfleet–Ockendon area, Corbets Tey Member 38
  Shepperton Gravel 102
  Silt Complex 98

Radiocarbon dating
  palaeogeographical evolution of Lower Thames Valley 175
Rainham Marshes 104
  Tilbury biogenic beds 115–16
Rainham–Orsett
  A13 borehole data 22–3
Ravensbourne, river
  confluence, Deptford 102
  Kempton Park Gravel Member 92
Reading Town Gravel
  correlations 196, 197
Red House Pit 20
Redbridge
  Fairlop–Rush Green spread 41
  Roding Valley Gravel 112
Redbridge Gravel 39, 192
Redbridge–Hackney
  East Cross Route, sections 42–3, 94
References
  Lower Thames Valley 201–13
Rhaxella chert 121
  Dartford Heath Gravel 23
  Great Monk Wood 18
  Richmond Hill Gravel 18
Rickson's Pit, Swanscombe Member deposits 25–6
Rivers
  braided 125–6
  sinuosity 125
Robin Hood's Pool Gravel 18, 179
Rochford Gravel 199
Roding, river 54, 112
  confluence, East Ham 104
  Ilford sediments 98
  Ipswichian 191
  origin 179, 180
Roding, valley 16, 18
  M11 56
Roding–Lea interfluve 39, 41
Roding–Lea–Thames confluence 39
Rom–Roding interfluve,

  Fairlop–Gants Hill 32
  Terrace fragments 94
Roding Valley Gravel 41
  gradient 112
  and Uphall Pit 80
Romford Road
  borehole data 56
Rotherhithe
  Surrey Docks 116–17
Royal Albert and Royal Victoria Docks
  Tilbury 116
'Rubble chalk' 126–7
Rutter's Pit, Crayford 69
  artefacts 171

St Paul's
  Mucking Gravel 56
St Swithin's
  Fairlop–Rush Green spread 41
Sand and fine sediments
  sedimentary structures 127–30
Sandy Lane Quarry *see* Aveley
Schmidt net 12
Sedimentary structures 123–30
  gravel units 123–6
  mass flow deposits 126–7
  sand and fine sediments 127–30
  standard symbols 13
Seven Kings Station
  Ilford 79, 80
Seven Kings Water
  Ipswichian 187
Shacklewell Lane
  Stoke Newington 80–1
Shadwell
  Mucking Gravel 56
Shadwell–Blackfriars
  borehole data 49
Shepperton Gravel 90, 100–7
  correlations 196
  defined 101
  Devensian 193
  distribution 101
  Flandrian 193
  long profile 108
  sites
    Blackwall 102
    Charlton, new 102
    Lea confluence 116
    Limehouse Reach 102
    London Docks 102
    Poplar Fire Station 102
    Purfleet 102
    Shepperton Quarry 101
    Tilbury Deposits Member 115–16
    Tilbury Docks 106
  thinning 104
  vertebrate faunas 135
Shepperton Member 54
Shooter's Hill Gravel 14–15
Shoreditch
  Clerkenwell–Shoreditch deposits 43, 48
Shrewsbury Lane Fire Station 14

Silt Complex
  absence/reduction in Central London 97–8
  Devensian 192
Slade Green
  brickearth 52
  Howbury Pit 89
Snaresbrook Park
  Corbets Tey Gravel 41
Sockett's Heath 30–1
South Woodford Gravel 94
Southern Outfall Sewer
  Plumstead 104
Spring Gardens Gravel
  and 'brickearth' 52
  Lower Thames 185, 196
  Trafalgar Square 52, 62
Staines alluvial deposits–Tilbury Deposits 114–17
Stamford Hill Gravel 49, 58
  Stoke Newington Common 82
Stanford-le-Hope
  Corbets Tey Gravel 39
Stepney
  Mucking Gravel 56
Stifford, North
  Orsett Heath Member 38
Stifford, South 64
  Orsett Heath spread 31–2
  West Thurrock Gravel 106
Stifford, South St Quarry
  Silt Complex particle-size 38
Stoke Newington Common 58, 80–6
  artefacts 172
  Bayston Road 85
  borehole data 47, 82–3
  Cazenove Road 82
  Church Street 83
  High Street 85
  Highbury Silts and Sands Member 83
  Northwold Road 82
  Rossington Street 82
  Shacklewell Lane 80–1
  Silt Complex 98
  solifluction 85
  Stamford Hill 82
  Stoke Newington Sand 85
  to Kingsland, section 58
  see also Hackney Downs
Stone Cross
  Orsett Heath Member 20, 31
Stone Marshes
  East Tilbury Marshes Gravel 90
Stoneham's Pit, Crayford
  artefacts 171
Stoneham's Pit, Crayford–Erith area 69
Stratford
  borehole data 56
  High Street 111, 117
Stubbers Youth Camp
  Ockendon, North 75
Surrey Docks
  Rotherhithe 116–17

Sutton Place
  Darent Valley 113
Swanscombe
  Mollusca 25
Swanscombe Member deposits 18–19, 24–9
  artefacts 24, 167–8
  Barnfield Pit 25–6
  Crayland's Lane 26–7, 28–9, 31
  geochronology 182
  Knockhall Cutting 99
  Lower Gravel 25–6
  Lower Gravel bed 181
  Lower Loam 26
  Middle and Upper Gravel 26–9
  principal components analysis (PCA) 120–2
  Rickson's Pit 25–6
  sequence summary 29
  vertebrate faunas 132–3, 136–7

Talbot's Pit, Crayford–Erith area 69
Taplow Gravel
  Mucking Member 43, 56
Taplow Terrace 43
  origins 49–51
Taplow-equivalent Mucking Gravel 43
  Corbets Tey Member 51
  see also East Tilbury Marshes Gravel
Terrace nomenclature 9–10
Terrace stratigraphy 8–9
Tertiary flint 121
Thames, river
  course 180
  palaeocurrent measurements 39
  see also Lower Thames Valley
Thanet Beds
  solifluction 62
Thanet Sand 30–2
  Crayford–Erith area 69
  gravel beds 19–20, 62
  M25 38
Third Terrace
  Lower Thames valley 49
Thurrock
  anticline see Purfleet
  East Tilbury Marshes Gravel 92
  see also Grays Thurrock
Thurrock Ridge 38
Thurrock, West 64
  artefacts 171
  Devensian Stage 191–2
  Elephas primigenius fauna 171
  and Grays area 52
  Hollin's sites
    WT1 66
    WT2 66
    WT3 64, 66
    WTA 67–8
    WTA and WTF pollen counts 143, 144
    WTD 66
  Ipswichian 187
  Lion Tramway Cutting 54, 64

  WTA and WTF pollen counts 143
  Mollusca 68
  palaeobotany 143–5
  particle-size distribution 67
  pollen analysis 66
  sections 65
  vertebrate faunas 68, 132–3, 138
  West Thurrock Marshes 106
  see also Aveley–West Thurrock–Crayford silts and sands; Gray Thurrock
Thurrock, West Thurrock Gravel Member 64–8, 68, 89
  brickearth 64
  Lion Tramway 89
  long profile 92
  Tilbury Deposits 89
Tilbury
  Docks 116
    Shepperton Gravel 106
  Limehouse Reach 102, 117
  Marshes, deposits 56
  River Thames, section, borehole data 91
  West, Orsett Heath Gravel 31
  World's End 115
Tilbury Deposits Member 54, 89, 114–17
  at East Tilbury 106
  Crossness 116
  deposition 194
  Isle of Dogs 116
  long profile 108
  Shepperton Gravel 115–16
  and Staines alluvial deposits 114–17
  Tilbury biogenic beds 115
  see also East Tilbury Marshes Gravel
Tottenham
  Lea Valley terraces 110
  section 111
Tottenham Hale
  Lea Valley Gravel 118
Tower Hamlets
  East Tilbury Gravel 56
  Silt Complex 97
Trafalgar Square
  Middle Thames 190, 196
  Spring Gardens Gravel 52, 62

Uphall Pit
  Corbicula bed 77
  and Roding valley 80
  sequence, Ilford 77
Upminster
  Cranham Hall 76
  Park Corner Farm 76
Upper Floodplain see Kempton Park Gravel
Upper Loam
  see also 'Brickearth'
Upper Loam facies 27–8, 61
Upper Middle Gravel unit 27–8
Upton Park
  Barking area 55

INDEX

Vertebral faunal assemblages 131–9
    cold environment 135–6
    preservation 134–5
    table 132–3
    taphonomy 134–5
    temperate environment 136–9
Victoria Line
    Lea Valley Gravel 111, 118

Walbrook, river
    confluence, London Bridge 102
    valley, London Wall 117
Waltham Forest 14
Walthamstow Reservoirs
    Lea Valley Gravel 118
Walther's Law 129
Wandle, river 179
    valley 14, 15–16
Wansant Pit deposits 18–19

Wanstead
    Corbets Tey Gravel 41, 49
    Lea–Roding interfluve 32
    Station borehole data 41, 43
    Wanstead Flats, terrace 55, 56
Warley Gravel 16
    palaeogeographical evolution 178
Waterloo Station 117
Weald
    Lower Cretaceous 122
West Ham
    Lea–Thames confluence 117–18
    Northern Outfall Sewer 111
West Thurrock *see* Thurrock West
West Thurrock–Crayford silts *see*
    Aveley–West Thurrock–
    Crayford silts and sands
Westmill Gravel 197
Wey valley 179, 180, 197

Whipps Cross
    Corbets Tey Gravel 49
Windsor–Chertsey
    Taplow Terrace 51
Winston Way
    Ilford 54, 55, 79–80
Wolstonian Stage 30, 183–5
    Hoxnian–early Wolstonian Stages 181–3
Woodford
    South
        M11 112
        Roding Valley Gravel 112
        South Woodford Gravel 94
Woodford Green 14
Woodford Green Gravel Member 179, 197
Woolwich beds
    Crayford–Erith area 70–1